D0154167

Islands of Truth
A MATHEMATICAL MYSTERY CRUISE

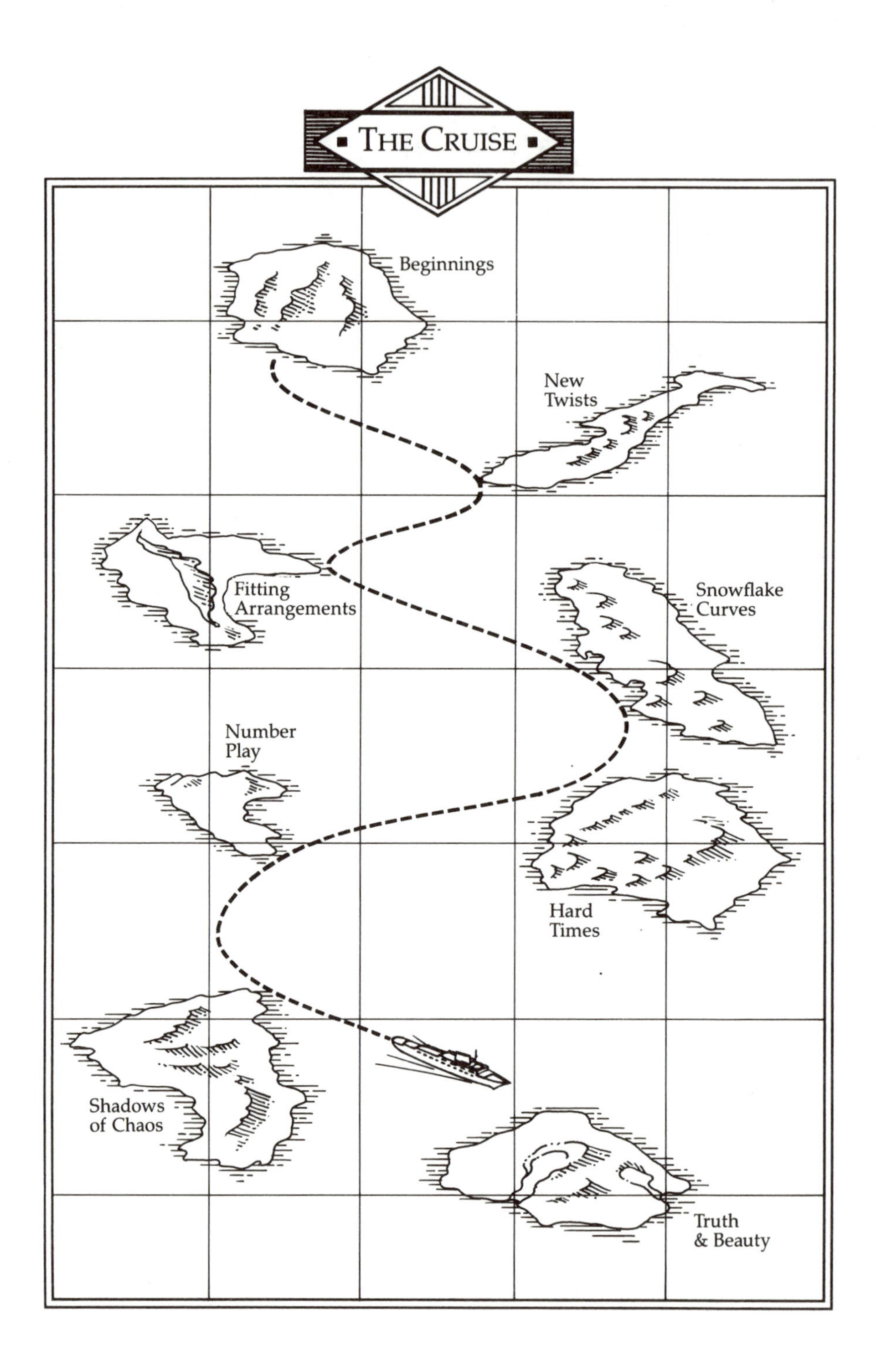
THE CRUISE
Beginnings
New Twists
Fitting Arrangements
Snowflake Curves
Number Play
Hard Times
Shadows of Chaos
Truth & Beauty

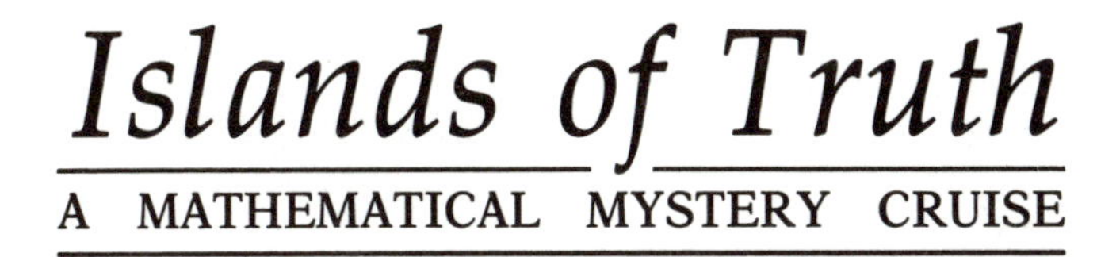

Islands of Truth

A MATHEMATICAL MYSTERY CRUISE

Ivars Peterson

W. H. Freeman and Company
New York

Library of Congress Cataloging-in-Publication Data

Peterson, Ivars.
Islands of truth : a mathematical mystery cruise
by Ivars Peterson.
p. cm.
Includes bibliographical references.
ISBN 0-7167-2113-9
1. Mathematics—Popular works. I. Title.
QA93.P474 1990 89-49501
510—dc20 CIP

Printed in the United States of America

1 2 3 4 5 6 7 8 9 0 VB 9 9 8 7 6 5 4 3 2 1 0

To our son, Kenneth Alan
Welcome to life's mysteries

Contents

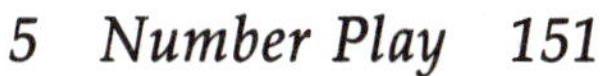

5 Number Play 151

6 Hard Times 194

Portraits of Mathematics

COLOR PLATE 1 Saddle Grid
Scherk's surface, discovered in 1835, looks like a checkerboard of saddle shapes. An example of a doubly periodic surface of constant mean curvature, it can be understood as the smooth joining of two sets of planes, one set at right angles to the other.

COLOR PLATE 2 Etruscan Venus
Studying a four-dimensional mathematical object by viewing its projection into three dimensions is like working out how a chair is put together with only its shadow as a guide. The Etruscan Venus, a geometrical surface that happens to have only one side, can be thought of as the three-dimensional "shadow" cast by a four-dimensional form topologically related to the Klein bottle.

COLOR PLATE 3 Ida Thoughts
A set of equations governing the motion and shape of an oval sweeps out a novel topological form now dubbed "Ida." Each colored section represents a piece of the surface generated by ovals moving through a particular range of angles.

COLOR PLATE 4 A Ring of Horses
Once mathematicians proved it mathematically possible to turn a sphere inside out without introducing creases, the problem became one of visualizing the process. In Bernard Morin's version of such an eversion, following the positions of only 12 points on a sphere is sufficient to define the entire transformation. The crucial moves occur when the 12 points define a polyhedron (seen from two different viewpoints) consisting of four nonconvex pentagons (termed "horses") and eight assorted triangles.

COLOR PLATE 5 The Heart of a Sphere
One way to program a computer to show a sphere eversion is to build the required surfaces from a patchwork quilt of mathematical expressions, each one defining a small piece of a particular surface. This computed figure, sliced open to give a view of its interior, reveals the sphere's contorted appearance at one point during the eversion.

COLOR PLATE 6 Knot Math
A mathematical knot, such as this trefoil knot, forms a complete loop, having no beginning and no end. To a topologist, a knot is formally any simple closed curve embedded in three-dimensional space.

COLOR PLATE 7 Pentagon Bees
In the 1970s, amateur mathematician Marjorie Rice discovered several new ways of fitting together pentagons to tile the plane. She used her pentagonal tilings as the basis for colorful, pleasing designs.

COLOR PLATE 8 Growing to Infinity
By fitting together Penrose tiles (fat and skinny diamonds) according to certain rules, it is possible to form a defect-free quasiperiodic tiling. Starting with a small "seed," the pattern grows in stages, going from one polygonal "dead surface" to another.

COLOR PLATE 9 Wintry Forest
Although the rules for generating this computer-generated image, based on a photograph of Germany's Black Forest, are simple, the picture shows a great deal of detail.

COLOR PLATE 10 Snow Carpets
Simple rule-based growth models can generate fractal patterns that look too regular to mimic real snowflakes. Modifying the rules to include an element of randomness produces more realistic pictures.

COLOR PLATE 11 Playing with Fire
Simple computer models can mimic the behavior of forest fires. In this snapshot from such a forest-fire simulation, yellow sites represent fires and green sites represent living trees. At each time step, new trees grow, other trees catch fire, and burnt trees die off.

COLOR PLATE 12 Pi-Scape
A computer can generate a fractal landscape representing the distribution of the first million decimal digits of pi, in which each digit is a step in a random-walk process.

COLOR PLATE 13 Pendulum Spaghetti
A spherical pendulum can wind up in any one of several different kinds of motion. This computer-generated image shows initial pendulum posi-

tions plotted against velocities. Orbits starting at points in the red regions are attracted to a periodic orbit that winds clockwise, whereas blue regions represent counterclockwise motion. The green and yellow regions designate initial points leading to two other types of motion.

COLOR PLATE 14 A Chaotic Question
This picture, known as the Ikeda map, is the result of substituting an initial value into a mathematical expression modeling the action of an optical switch, computing the answer, then substituting that answer back into the expression, and so on.

COLOR PLATE 15 San Marco Wonderland
The transcendental function known as the hyperbolic cosine has many curious properties. In this representation, the lobed red shape in the middle of the figure is called the San Marco curve. Miniature copies of the same curve surround the body.

COLOR PLATE 16 Four-Dimensional Addition
Quaternions may be considered as extensions of the notion of "number" to four dimensions, with the rules of quaternion multiplication and addition providing a way of doing arithmetic on four-dimensional quantities. Such mathematical constructs can be visualized as strange, contorted, three-dimensional shapes.

Preface

"What is the use of a book," thought Alice,
"without pictures or conversation."
—Lewis Carroll, *Alice's Adventures in Wonderland*

My earliest recollection of anything mathematical is of my struggle, several years before grade school, to remember that "five" goes between "four" and "six." As far as I could tell, learning the names of numbers was nothing more than memorizing a set of arbitrary labels in their correct order. The deeper meaning of numbers and their connection with counting eluded me.

That was the first of many mathematical mysteries I encountered over the years. Usually, they weren't genuine mysteries. They were just words, ideas, or rules that puzzled me for a while. In time, I figured them out, and my understanding of things mathematical advanced by small, hard-won steps.

I can look back at the notes I took one summer in the middle of my high-school years when I was lucky enough to spend six weeks in a special science program. The notes record my first serious encounters with topics such as group theory, linear programming, and quantum mechanics. The mathematics we did then seems simple now, but I still remember how difficult it was for me to grasp what was going on. It felt as if I were witnessing expert conjurers at work. I could admire their skill, but I didn't understand how the tricks were done.

Later, as a math and science teacher, I enjoyed working with high-school students, helping them over hurdles that I recalled having run up against myself. I particularly liked the chance to revisit and become more familiar with ideas in science and mathematics that, at first, had seemed so puzzling to me. It was a pleasure to come back to problems that had once seemed impossible and to find that the route to the answers was clear and straightforward.

My more recent experiences as a journalist reporting advances in mathematics and science have considerably broadened my view of mathematics. There's still a great deal of mathematics I know little about and barely understand. But the glimpses of mathematical research I get are fascinating and, in this age of computer graphics, they're made even more striking by sophisticated pictorial images. What I have come to appreciate is how much the understanding of mathematics requires hard, concentrated work. It combines the learning of a new language and the rigor of logical thinking, with little room for error.

I've also learned that mystery is an inescapable ingredient of mathematics. Mathematics is full of unanswered questions, which far outnumber known theorems and results. It's the nature of mathematics to pose more problems that it can solve. Indeed, mathematics itself may be built on small islands of truth comprising the pieces of mathematics that can be validated by relatively short proofs. All else is speculation.

My goal in this book is to share some of the mathematical mysteries now at the frontiers of research. You don't have to be a mathematical insider to appreciate them.

One of the greatest mysteries is why mathematics seems to be the ideal language for describing and explaining natural phenomena. Over and over again, scientists find that a mathematical structure discovered years before with no practical application in mind turns out to be a useful way of formulating a scientific theory to describe or account for some recently observed phenomenon.

The interactions of computation, science (particularly physics), and mathematics make an intriguing story full of surprises: studying how one mathematical shape can be transformed smoothly into another leads to new forms of computer-generated art; tallying all possible ways of arranging geometric shapes provides a step on the road to computers as architects; investigating fractals—geometric shapes that repeat themselves on many different scales—leads to models of mountains and clouds. Number theory, once considered the purest area of mathematics, now has applications in computer science and acoustics. These are just a few of the topics you'll meet in this volume.

Much of the material in this book has appeared in a somewhat different form in *Science News* over the last 8 years. I'm grateful to *Science News* editor Patrick Young for encouraging my pursuit of mathematical truth and for his insistence that what I write be understandable to as broad an audience as possible.

Because I am not a mathematician myself, I have relied on the advice, comments, and writings of many mathematicians. Lynn Arthur Steen and Ron Graham have been especially helpful in pinpointing significant developments and patiently explaining their relevance and importance. Graham's remarks on the difficulty of establishing mathematical certainty inspired the title of this book. Steen's essay "The Science of Patterns" provided an important theme.

The bibliography lists many of the other sources for this book. Two essays in particular played key roles: physicist David Gross's witty look at the relationship between physics and mathematics, and mathematical physicist David Ruelle's ruminations on mathematics as a human construct. Unfortunately, the bibliography can't list all the interviews, conversations, discussions, debates, and lectures that provided important clues toward unraveling the mathematical mysteries described here.

I wish to thank all the mathematicians and scientists who helped me put this book together, by explaining ideas, providing illustrations, and supplying reference materials. Any errors, however, are my responsibility.

I'm grateful to my wife Nancy, who was the first to read these pages, for many helpful suggestions. I also appreciate the efforts of my father, Arnis Peterson, who translated my rough drawings into clear diagrams, and Phil Hodge, who had a chance to play games on his computer to come up with several of the stranger illustrations I needed for this book. Again, the team at W. H. Freeman did a wonderful job in transforming my stack of loose pages, sketches, photographs, and illustrations into a beautiful, finished book.

Ivars Peterson
January 1990

Islands of Truth
A MATHEMATICAL MYSTERY CRUISE

Beginnings

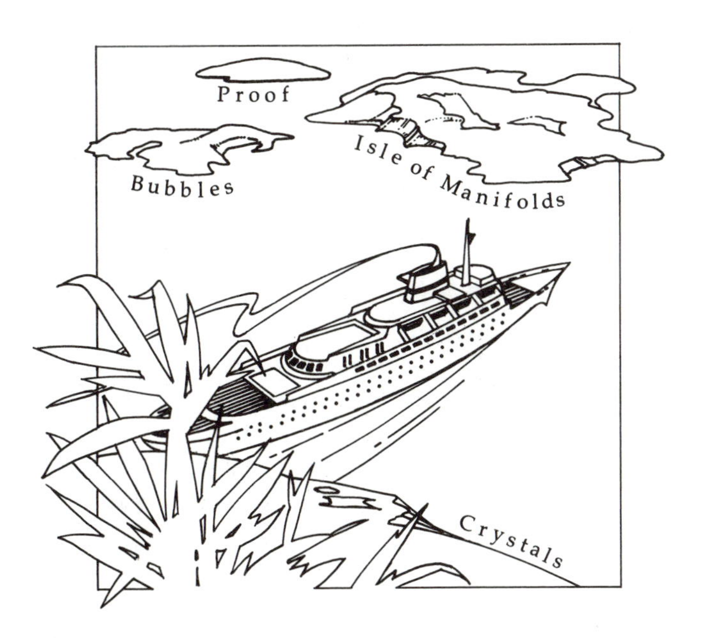

One, two, three, four. . . .

We learn to count at such an early age that we tend to take the notion of abstract numbers for granted. We know the word "two" and the symbol "2" express a quantity that can be attached to apples, oranges, and other objects. We readily forget the mental leap required to go from counting specific things such as apples to the abstract concept of number as an expression of quantity.

Abstract numbers are the product of a long, slow cultural evolution. They rank among the greatest of human contrivances. Studies of the development of mathematics often herald the invention of zero and the advent of place notation as major accomplishments, while mistakenly assuming that abstract numbers are intuitive.

Mathematician and philosopher Bertrand Russell wrote in his book *Introduction to Mathematical Philosophy*: "It must have required many ages to discover that a brace of pheasants and a couple of days were both instances of the number 2." Evidence supporting that statement is now surfacing in archaeological studies of an ancient Mesopotamian counting system that gave rise to the idea of abstract numbers.

Tokens of Plenty

Archaeologist Denise Schmandt-Besserat didn't start out with abstract numbers in mind. She was looking for the earliest examples of the human use of clay when she started her quest two decades ago. She went from museum to museum, reviewing Middle Eastern clay collections from cultures that thrived between 10,000 and 6000 B.C. She expected to see bricks and pots. Instead, she was surprised to discover hoards of little clay objects that could easily pass for children's playthings.

Excavators had recovered the baked-clay objects from archaeological sites ranging from Turkey and Palestine to Syria and Iran. In some places, they had found only a few specimens. Occasionally they had encountered large collections of these mysterious items. The objects came in a variety of geometric shapes, including cones, spheres, disks, cylinders, and pyramidlike tetrahedrons. Some appeared to be miniature models, an inch or less in size, of animals, tools, and other natural or human-made items. Other objects were perforated and marked by various patterns and lines.

Traditionally, archaeologists had categorized such objects by shape and tried to guess what a particular class of objects would have been used for. The disks, they thought, may have been lids for small jars. The spheres could have been marbles.

Schmandt-Besserat, however, looked at what the objects had in common. They were all made of clay, similar in size, and manufactured in roughly the same way. To her, it was obvious the objects belonged together. Once they were considered as a group, their role in counting and record keeping gradually became evident.

First, Schmandt-Besserat realized that she was dealing with the precursors of writing. She tracked symbols written on clay tablets, moving backward in time from what are known as cuneiform markings to pictographs to object shapes. In a few cases, she managed to determine the meaning of specific objects. After years of study, she saw how the same objects, or tokens, were also connected with counting and numbers. Her research tells a remarkable story of the stages whereby human culture at first slowly and then ever more rapidly mastered the art of abstraction.

The first appearance of clay tokens in the archaeological record of the Middle East coincides with the development of agriculture, particularly grain cultivation, in the period from 8000 to 7500 B.C. The Sumerians, formerly hunters and gatherers, began settling in villages in the fertile valleys of the Tigris and Euphrates rivers.

Archaeological studies of the period show evidence of grain cultivation in fields surrounding villages, the construction of communal silos for storing grain, and a rapid increase in population. In such a setting, individual farmers needed a reliable way to keep track of their goods, especially the amount of grain stored in shared facilities. They began to fashion clay tokens into simple geometric shapes with plain, unmarked surfaces (see Figure 1.1).

Sumerian farmers kept track of their goods by maintaining stores of baked-clay tokens—one token for each item, different shapes for different types of items. A marble-sized clay sphere stood for a bushel of grain, a cylinder for an animal, an egg-shaped token for a jar of oil. A cone represented an amount of grain roughly equivalent to a liter, whereas flat disks probably signified larger amounts. Two clay spheres would have stood for two bushels of grain, and two egg-shaped tokens for two jars of sesame oil. There were as many tokens, or counters, of a certain shape as there were of that item in the farmer's store.

The representation of real goods in a one-to-one correspondence meant that tokens could be lined up in front of accountants, who doubtless organized them according not only to types of goods, but also to types of transactions, and earnings or expenditures by producers or recipients. The tokens could even be arranged in visual patterns to make estimation and counting easier. Still, the concept of "two" or of any other number, as a quantity, didn't exist.

This simple system of data storage persisted practically unchanged for almost 4,000 years, spreading over a large geographical area. Eventually the growth of villages into cities and the

Figure 1.1 Plain clay tokens in the shape of cones, spheres, disks, tetrahedrons, and cylinders were used in ancient Mesopotamia before the invention of writing to keep track of goods. Each token was a symbol representing one unit of a particular commodity. The cones, spheres, and disks probably stood for different measures of grain, whereas the tetrahedrons may have represented units of labor and the twisted cylinder a bundle of rope.

increasing complexity of human activities, especially in southern Mesopotamia, forced a shift to a more versatile means of record keeping. That shift was marked by the appearance of elaborate tokens alongside the well-established system of simple counters. Though similar in size, material, and color and fabricated in much the same way as their plainer cousins, the new tokens bore various kinds of surface markings and showed a greater variety of shapes. There were twin cones, bent coils, and miniature models of tools, utensils, containers, and various animals (see Figure 1.2).

The elaborate tokens were used for manufactured products —the output of Sumerian workshops. Incised cones and diamondlike rhomboids probably represented loaves of bread and vessels of beer. Disks and parabolic tokens with linear markings signified different types of fibers, cloths, and finished garments. Incised cylinders and rectangles stood for strings and mats. Other tokens seem to have represented luxury goods, including perfumes and various kinds of metalwork.

The advent of complex tokens coincided with the rise of powerful central governments and the construction of monuments and great temples, beginning around 3350 B.C. Art from that period shows the rise of a governing elite and the pooling of community resources for celebrating large festivals.

The management of massive construction projects and public events required considerable planning and the bringing together of vast quantities of goods. The token system, now extended to cover goods and services, played a key role in managing such tasks. It was precise enough for keeping comprehensive records,

Figure 1.2 Complex tokens had a variety of shapes and often bore markings or were perforated. The disk with a set of parallel stripes probably represented a ewe.

providing the leadership with an effective means of exercising control over the populace. Indeed, scenes from Sumerian art depicting beatings and other punishments for tax evasion suggest that the practice was common.

Throughout this period of increasing urbanization, simple tokens associated with products of the granary and farm coexisted with complex tokens representing products of the urban workshop. In fact, temple excavations reveal that the Sumerians handled the two kinds of tokens separately and stored them differently.

Sets of simple tokens were kept in clay globes, or "envelopes," that often bore markings indicating what was enclosed and seals that may have recorded transactions (perhaps contracts, receipts, or even IOUs). Temple clerks marked the envelopes simply by pressing tokens into the soft clay before sealing and baking the globes, making visible the number and shape of tokens enclosed. Excavated specimens show circular imprints left by spheres and wedge-shaped imprints left by cones.

Once sealed in their clay cocoons, the tokens were hidden from view. It didn't take long for busy bureaucrats to realize that once the clay envelopes were marked, it was no longer necessary to keep the tokens. In fact, the marks by themselves, impressed on a clay tablet, were sufficient.

Complex tokens couldn't be stored in clay envelopes because they left indecipherable impressions. Instead, perforations allowed such tokens to be strung together, with special clay tags apparently identifying the accounts. In this case, the shortcut the bureaucrats discovered was to inscribe the incised pattern found on the surface of a complex token directly onto a clay tablet. For example, an incised ovoid token could be replaced by a neatly drawn oval with a slash across it. Such pictographic signs would be sufficient to indicate the nature of the items being recorded.

The result was a new, more practical, less cumbersome data storage system. A small set of clay tablets with neatly aligned signs was much easier to handle than an equivalent collection of loose tokens. And using a stylus for marking clay tablets was a lot faster than making an impression of every token.

Then came a leap so great that the accountant who performed it might well be called *homo mathematicus*. Around 3100 B.C., someone had the bright idea that, instead of representing 33 items, say, jars of oil, by repeating the symbol for one jar 33 times, it would be simpler to precede the symbol for a jar of oil by numerals—special signs expressing numbers (see Figure 1.3). Moreover, the same signs could be used to represent the same quantity of any item.

The signs chosen for this new role were the symbols for the two basic measures of grain. The impressed wedge (cone) came to stand for 1 and the impressed circle (sphere) for 10. There were no special symbols for other numbers.

In this way, the token system evolved into a kind of shorthand in which signs indicating standard measures of grain, impressed

Figure 1.3 This pictographic clay tablet from a site in present-day Iran carries the symbols for the numeral 33 (three wedges and three circles) and the sign for a jar of oil. The inscription can be read as "33 jars of oil."

on a clay tablet, came to represent not grain or any other specific commodity but the concept of pure quantity. The coupling of signs for numbers, or numerals, with pictorial symbols for specific goods provided ancient accountants with the tools they needed to record the multitudinous and varied transactions of a city.

It was a revolution in both accounting and human communication. For the first time, there was a reckoning system applicable to any and every item under the sun. It put an end to a cumbersome scheme in which different tokens had been used for counting different goods. It also made taxation on a broad scale feasible. Anything of importance could be expressed compactly and flexibly on clay tablets.

This economical notation spread rapidly. Clay tokens themselves became obsolete by 3000 B.C., replaced by pictographic tablets that could represent not only "how many" but also "where, when, and how." With the introduction of a new type of stylus, pictographic writing developed into cuneiform notation. The resulting record-keeping system was so efficient and convenient that it was used in the Middle East for the next 3,000 years. Eventually, it was displaced by Aramaic script written with a flowing hand on papyrus, which proved even more efficient.

The Sumerian method for recording quantities was just the first step on the long road toward the development of an abstract number system. What started as merely a superior accounting scheme quickly led in later civilizations to further mathematical innovations and deeper levels of abstraction.

The Babylonians adopted and extended the Sumerian system, introducing a form of place notation, in which symbols took on values depending on their relative positions in the representation of a number. They even developed the equivalent of today's decimal notation, making it just as easy to add and multiply fractions as it was to add and multiply whole numbers. Babylonian mathematicians were also successful in finding procedures, or algorithms, for computing quantities such as square roots. These ideas and other Babylonian concepts in algebra and geometry eventually traveled to Greece, at length becoming part of the great flowering of mathematics in the ideas of Pythagoras, Euclid, and Archimedes.

Whether a similar step-by-step process leading from concrete counting to the concept of abstract numbers occurred independently in other civilizations, such as the Egyptian, Chinese, or Mayan, isn't known. No one has taken a serious look at the development of counting in such settings to see how the transition to abstract expressions of quantity occurred. Like most people, archaeologists and cultural anthropologists mistakenly tend to assume that abstract counting is a given.

The English language, for one, still contains traces of concrete counting—the use of different terms for counting different things. Terms such as "couple," "twin," "brace," and "pair" all mean "two" but are often used in different contexts: a pair of shoes, a brace of pheasants. Our language also features different collective nouns for different groups of animals: a school of fish, a flock of sheep, a herd of cows, a pride of lions. These present-day vestiges of concrete counting make the Sumerian achievement thousands of years ago all the more remarkable.

New Math

Students, faced with the chore of memorizing multiplication tables, struggling through geometric proofs, or pondering how long it takes a slowly leaking conical vessel to drain, often have the feeling that mathematics is little more than an unchanging body of knowledge that must be painstakingly and painfully passed on from generation to generation. Missing is a sense of how mathematics has evolved since its origins in the distant past and how new mathematics is constantly being discovered and created.

Even scientists and engineers often harbor an image of mathematics as a well-stocked warehouse from which to select ready-to-use formulas, theorems, and results to advance their own theories. Mathematicians, on the other hand, see their field as a rapidly growing enterprise furnishing a rich and ever-changing variety of abstract notions.

The development of calculus is a good example. It's hard to imagine that something now as fundamental and pervasive as calculus—the bane of many a beginning college student—didn't exist three hundred years ago. Isaac Newton invented the basic concepts underlying calculus to describe more effectively how the positions of bodies change over time, especially the way the force of gravity acts on planets or on a falling apple. Later mathematicians elaborated and refined Newton's calculus, establishing the foundations of what is now known as analysis. Only a century ago, James Clerk Maxwell used Newton's mathematical language to write out the laws of electromagnetism. Early in the nineteenth century, Bernhard Riemann applied similar ideas to geometry, setting the stage for Albert Einstein, who discovered in Riemannian geometry the key to a modern theory of gravitation.

Mathematical discovery begins with the study of examples. Mathematicians gather information in the form of lists of numbers, diagrams of knots, sketches of geometric figures, the results of computations, or even observations of physical phenomena. Going through a process of abstraction, they distill from these examples what they believe are essential, common features. The mathematical concept of number is an abstraction of the process of counting; the concept of space is an abstraction of our own physical experiences in a three-dimensional world. Similarly, the mathematical notion of a function represents a formal way of looking at human ideas of measurement and motion, and probability is tied to experiences such as flipping a coin.

It is the process of abstraction, and the language that goes with it, that makes mathematical terrain so difficult for nonmathematicians to penetrate. Outsiders find it hard to get at the information encoded in mathematical formulas because little or nothing in the actual patterns of the stark symbols on a printed page offers them a clue to the formula's meaning. To an outsider, the manipulation of such symbols looks like a private, perhaps magical, game aimed toward mysterious, unworldly ends. Mathematicians also appropriate simple, everyday words for their own purposes, using them in unexpected ways or assigning to them specific, technical meanings to express abstract concepts. Many easy words stand for complex ideas: function, group, domain, root, imaginary, radical, derivative, transcendental, chaos, field, space, and so on. Guessing the mathematical meaning of any one of these words would be like guessing from its name that a drug store is a place to buy cameras, magazines, and car wax.

The word *function* serves as a particularly pertinent example because it is both a common word and the name for a key mathematical concept that comes up over and over again. In the realm of mathematics, a function is neither a social event nor a bodily process, or even a duty. Mathematically, a function expresses the relationship between two sets of numbers or other mathematical entities, establishing a specific correspondence between the two lists (see Figure 1.4). Scientists use the same concept when they prepare a table or plot a graph showing, say, how the temperature of a liquid changes with time or how the speed of sound depends on a material's density.

In general, the language used in mathematics is unusually dense. The position and precise meaning of every word and symbol make a difference. Moreover, mathematics seems to deal with familiar things in strange ways. For instance, mathematical knots, unlike everyday knots familiar to boy scouts and sailors, form closed loops with no loose ends. The words may be simple and commonplace, but in mathematics, the concepts they express are often deep and subtle.

Mathematicians are concerned with structures and relationships. They look for patterns. In fact, the motivation for much mathematical research is the desire to gain an understanding of the inherent structure of mathematics, to unearth previously hidden fragments that may link one piece of the structure to another.

Furthermore, although mathematics originated with the practices of counting and measuring, it actually deals with logical reasoning in which theorems—statements about the behavior of mathematical objects, whether numbers, lines, or knots—can be deduced from initial assumptions. In other words, there are for-

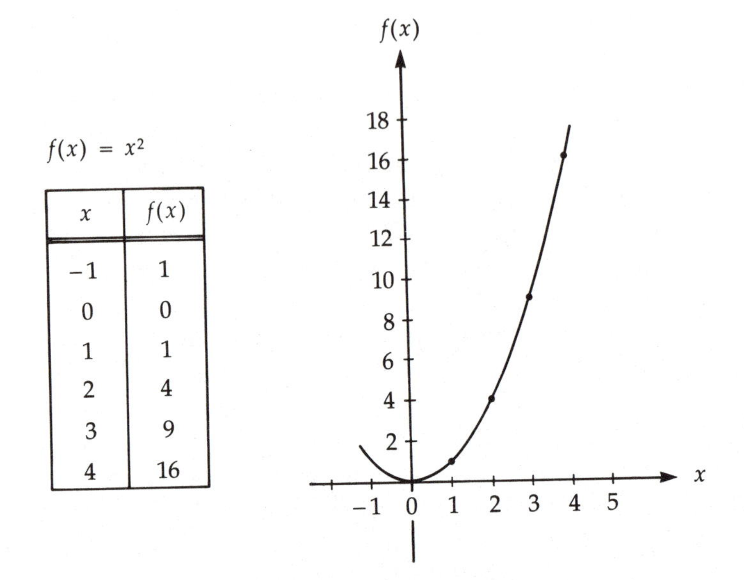

$f(x) = x^2$

x	$f(x)$
−1	1
0	0
1	1
2	4
3	9
4	16

Figure 1.4 An expression such as x^2 has no definite numerical value until we assign x a value. We say that the value of this expression is a function of the value of x, and write $f(x) = x^2$. Thus, when $x = 2$, $f(x) = 2^2 = 4$; when $x = 3, f(x) = 3^2 = 9$. Similarly, we can find by direct substitution the value of $f(x)$, whatever the function happens to be, for any number x. A function may also be expressed in the form of a table (*left*) or as a curve plotted on a graph (*right*). In this example, the curve is a parabola.

mal methods known as proofs to discover mathematical truths. In such proofs, mathematicians manipulate symbols according to fixed logical rules to arrive at true statements.

When the starting assumptions and symbols correspond to real things in the real world, the conclusions that come out of a proof say something about reality. For example, the set of abstract symbols "$2 + 2 = 4$" expresses a true mathematical statement. It also happens to mean, in the real world, that adding two apples to two apples is the same as adding two oranges to two oranges. In the same way, mathematical expressions known as differential equations provide uncannily precise means for describing physical phenomena such as a swinging pendulum and the propagation of light.

The insistence on absolute proof for determining the truth of statements differentiates mathematics from the sciences. In general, scientists are content to build their theories on the back of empirical data. The more convincing the data and the stronger the evidence, the happier they are with their theories. Scientists don't insist on a logical proof showing that nothing else could possibly be the cause.

Mathematicians, like philosophers, must step beyond the plausible and self-evident, striving for truth beyond the shadow of extreme doubt. For instance, no normal person is likely to deny that a simple closed curve (one that doesn't cross itself) on a flat surface, or plane, divides the plane into two regions: the inside and the outside (see Figure 1.5). Nevertheless, mathematicians were sufficiently skeptical that for many years they wondered whether every possible closed curve truly has just an inside and an outside. No satisfactory answer came until the nineteenth cen-

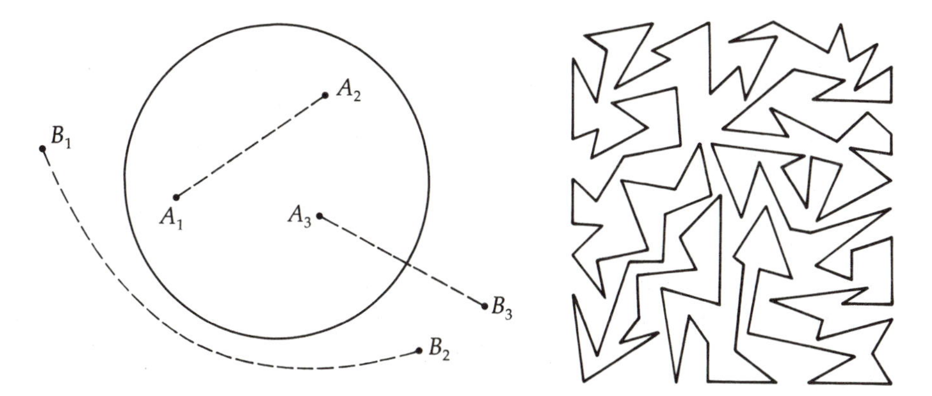

Figure 1.5 Proving mathematically that all simple closed curves have an inside and an outside is surprisingly difficult. In other words, such a loop, which doesn't cross itself, divides the plane into exactly two regions. That means each point of the plane falls into one of two classes: class **A** includes points that fall inside the loop, and class **B** includes points outside the loop. Any pair of points of the same class can be joined by a curve that doesn't cross the boundary, whereas any curve joining a pair of points belonging to different classes must cross the boundary. That's obviously true for a circle (*left*) but not so easy to see for a complicated curve like the polygon shown (*right*).

tury, when Camille Jordan stated the question explicitly and gave an intricate, lengthy proof that all simple closed curves in a plane truly have exactly two regions.

The reason for mathematical caution is that many times intuition or strong experimental evidence turns out to be misleading or wrong. The analog in three-dimensional space of Jordan's closed curve is the surface of a sphere, whether smooth or wrinkled. A balloon, for instance, clearly has just an inside and an outside. But mathematicians can imagine intricate, convoluted surfaces for which an inside and an outside can't be easily defined (see Figure 1.6).

Evidence based on a finite number of cases or examples isn't enough either, because the one exception may happen to lie beyond the range of experiment or computation. Attempts to settle the Riemann hypothesis, probably the most important unsolved problem in mathematics, serve as an illuminating example. That hypothesis makes predictions about the behavior of a certain set of numbers, and mathematicians have done calculations, most recently on high-speed computers, showing the hypothesis is true for the first 1.5 billion cases. But that's not sufficient to show it's always true.

The case of the so-called Mertens conjecture, once thought to be a possible path toward proving the Riemann hypothesis, reveals what can go wrong. This conjecture concerns the behavior of a certain function defined for positive whole numbers. Over the years since the conjecture was first proposed more than a century ago, mathematicians had been able to show that it was true for values up to 10 billion. But in 1984, Andrew Odlyzko and Herman te Riele took an indirect route and discovered that for a sufficiently large number the conjecture fails.

Odlyzko and te Riele used high-speed computers to find a kind of average for the sum required by the Mertens conjecture. Because an average is always less than the largest number in the set of values being averaged, it was enough to prove that the average itself was a sufficiently large number to disprove the conjecture. The researchers found such an average, although they did not come up with a specific number at which the conjecture is

Figure 1.6 Mathematicians sometimes invent bizarre objects to test their ideas. Alexander's horned sphere is one example of a convoluted, intertwined surface for which it is difficult to define an inside and an outside.

violated. Their best guess is that the number is bigger than 10^{30}, perhaps something as outrageously large as $10^{10^{70}}$—well beyond the range of any conceivable computer.

Absolute certainty is important because mathematics is constructed like a giant house of cards: one theorem piled on top of another, built upon a foundation of a small number of starting assumptions, or axioms. If one element were faulty, then the whole structure could come tumbling down. Fortunately, a network of internal checks and balances helps ensure consistency and proper connections. No single mathematician is ever likely to topple the entire enterprise.

Often, the obsession with proof is itself an important source of new ideas and mathematical methods. Efforts to prove that closed curves divide space into an outside and an inside led to the new mathematical field of algebraic topology, a central topic in modern mathematics. It's unlikely that any attack on a particular practical problem would have led to such novel abstract ideas.

Operating in a world of mental abstractions, mathematicians often pursue mathematics for its own sake, rather than to solve immediate problems or to achieve practical goals. Nonetheless, mathematicians regularly return to the natural and physical sciences for new problems and insights. Discoveries continue to originate from attempts to study and understand the physical world. More and more mathematicians are realizing that their discipline does not trickle down to applications but lives in partnership with the sciences, exchanging ideas, concepts, problems, and solutions.

And more quickly than ever, new mathematical ideas are slipping into applications covering wider and wider areas of science and technology. Mathematics is helping biologists track and predict how a molecule as complicated and convoluted as a protein manages to fold itself into a compact package. Economists use mathematical models to measure the economic factors contributing to a nation's competitiveness. Mathematical pursuits as abstract and esoteric as number theory and logic find roles to play in computer science and cryptography.

The availability of high-speed and high-capacity computers has been an enormously important force in helping to draw mathematics and the sciences closer together. Today's computers are bringing new areas of mathematics into existence, giving mathematicians new tools for solving problems. Computer science itself is generating a wealth of mathematical questions and prompting a fresh look at ways of tackling classical mathematical problems. Computer-based methods of constructing proofs are raising basic questions about what constitutes a mathematical proof. "Applications, computers, and mathematics form a tightly coupled system yielding results never before possible and ideas never before imagined," observes mathematician Lynn Arthur Steen.

Stretching Exercises

Pictures and physical models have long played an important role in mathematics. Euclid, more than two thousand years ago, founded his geometry on the rich imagery of lines, points, circles, and other regular forms. Many centuries later, Islamic artists and architects interested in symmetry for decorating their mosques discovered how rotations, reflections, and translations of variously shaped tiles combine to create breathtakingly beautiful patterns. The same complex interrelationships now find expression in the tenets of the mathematical subject known as group theory.

Leonardo da Vinci and other fifteenth-century painters, trying to solve the problem of representing spatial depth on a flat surface, spawned the topic of projective geometry. The artists noticed that parallel lines meet in vanishing points and proportions change according to strict rules. The same concepts are central features in the subject's modern mathematical counterpart.

Creative mathematicians testify to the importance of mental imagery in their work. The stark symbolic formulas of higher mathematics represent merely the final stage in a thought process more often than not beginning in the concrete and ending in the abstract. Moreover, we learn geometry by sketching triangles and curves. We teach the meaning of mathematics through drawings, constructions, and analogies to everyday experience.

Today's computer graphics—inhumanly fast, accurate, and consistent—provides a new tool for visualizing and exploring mathematical ideas. Incredibly versatile, computers can draw just about any shape that can be formulated as an algebraic expression. Translated into a step-by-step recipe, or algorithm, such a formula guides the computer's pen. Previously limited to a small repertoire of easy-to-draw curves, mathematicians can now visually explore forms represented by a vast array of mathematical expressions and formulas, from complex polynomials to fractals (patterns that repeat themselves on ever-smaller scales). They can seek patterns among forms never before visualized.

Consider a mass of soap bubbles—the kind of froth that can bubble up in a kitchen sink. At first glance, the basinful of iridescent bubbles seems nothing more than a haphazard collection of pliant, transparent balls randomly crowding one another. Nevertheless, a delicate balance holds sway throughout its ephemeral architecture. Mathematical rules specify the restricted number of ways in which bubbles cluster.

Observing soap bubbles raises questions about the rules that determine the natural, or optimal, shapes of other surfaces. What are the factors governing the geometry of optimal surfaces when mineral crystals or biological cells are packed together in complex configurations? Such a question inspires high-level mathematical research in which computer graphics plays an increasingly important role.

The surface of a glistening soap film tightly stretched across a closed wire loop, even a very knotted loop, is the smallest possible area spanning the loop. That minimal surface is a reflection of the soap film's tendency to seek a state of lowest energy. A soap bubble is spherical for the same reason. Any other closed shape of the same volume would have a higher surface energy.

Mathematician Jean Taylor has extended this idea to crystals. The key difference is that whereas soap films and soap bubbles have a uniform surface energy, crystals do not. Different crystal faces may have different surface energies. The shape of a single crystal is therefore not necessarily round like a single soap bubble. It usually has flat faces.

Using her recently developed theory, Taylor can compute and display the different types of minimal surfaces that a crystal surface assumes within a given boundary. These crystalline minimal surfaces are the solid analogs of soap films confined within a certain wire loop (see Figure 1.7). Taylor is using such images of minimal crystals to study how changes in the crystal shape affect the forms of corresponding minimal surfaces.

Taylor uses the computer as a tool for suggesting and testing conjectures about optimal surfaces. "There are a lot of phenomena and not very many theorems to explain them," Taylor says. For example, she has noticed that whenever a crystal's smoothly curved surface meets a flat surface, the corresponding minimal

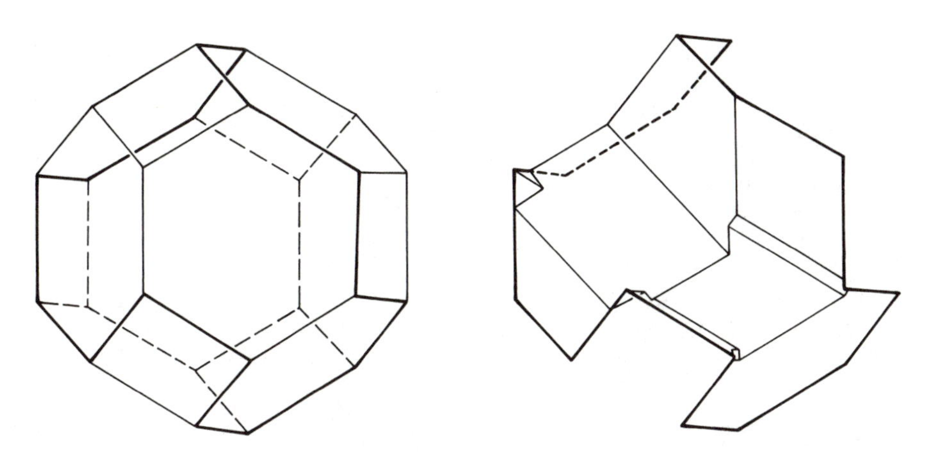

Figure 1.7 All surfaces in nature, including soap films and crystal faces, have a surface energy, which at equilibrium would be as small as possible. Here, a single crystal's equilibrium shape (*left*) sits beside one of the possible minimal surfaces (*right*) that fits within the boundary marked on the crystal. Together, this pair of crystal forms is the equivalent of a soap bubble next to a soap film stretched across a wire frame.

surface has a cusp (see Figure 1.8). Her computer experiments so far support this conjecture. Eventually, her studies may lead to a better understanding of why metals (which generally consist of interlocking crystalline grains) are strong, how they change form when heated, and when they might fail.

An important part of Taylor's research is the development of algorithms to compute and display optimal geometries for a variety of surfaces. The more complicated is the geometry that such methods can generate, the better the methods are for solving real-world problems.

Several years ago, engineers working on a rocket fuel tank encountered just such a problem. The tank held liquid fuel and had to operate in space. With the vehicle's motion virtually canceling out the effects of gravity, there was no guarantee that fuel wouldn't end up stranded on the side of the tank opposite to the fuel outlet. The engineering solution was to use capillary attraction, the same force that allows a sponge to soak up water. That

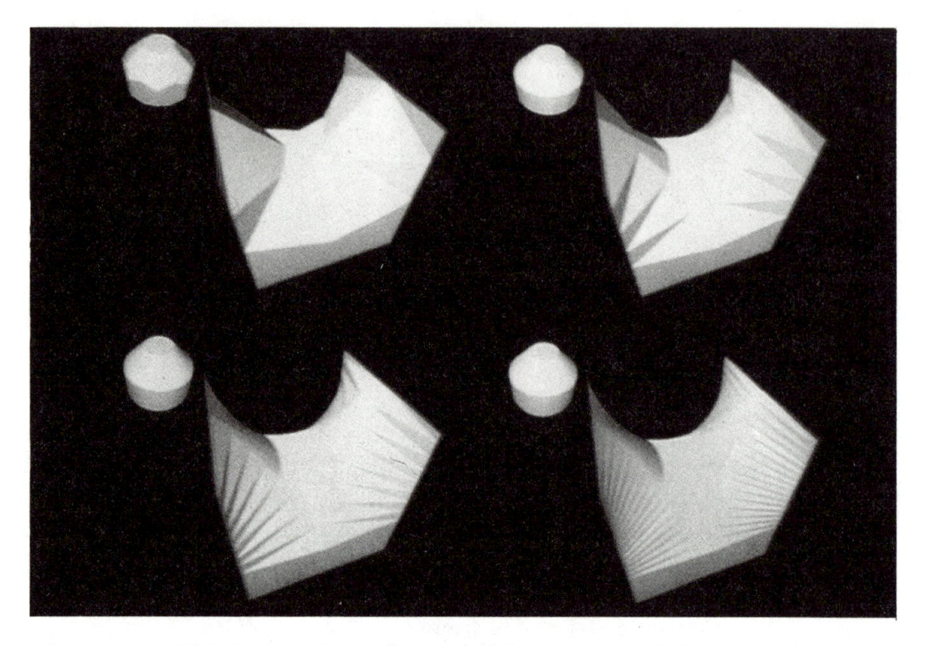

Figure 1.8 This image shows four equilibrium crystal shapes and a corresponding crystalline minimal surface for each. The equilibrium crystal shapes have 26, 50, 98, and 194 facets. The fourth one (*bottom right*) has so many facets, the crystal appears smoothly curved. Its minimal surface looks smooth in some places and corrugated in others. As the number of facets increases, the corrugations get finer but never quite smooth out. One can also see four cusped corners developing in the central facet. In some cases, the eye can see trends in features, such as the growth of cusps or the compression of corrugations to an infinitessimal width, that have not yet been proved mathematically to exist except in special cases.

meant filling the tank with a network of partitions to attract the fuel. The mathematical tools to find the optimal geometry for such a set of partitions are now starting to appear.

Mixing the polymers polystyrene, the stuff of plastic cups, and polyisoprene, a material used for making automobile tires, is like trying to combine oil and water. The two polymers repel each other. However, chemists can bond the polymers together to produce what is known as a block copolymer. Like belligerent but exhausted adversaries forced to attend a peace conference, the two materials are inextricably linked yet avoid contact with each other as much as possible. And like the shapes of soap films, the geometry of the interface of two such linked but repelling polymers is closely related to the mathematical behavior of minimal surfaces—surfaces that take up the least possible area within a certain boundary. That connection has led to an unusual collaboration between polymer physicist Edwin L. Thomas, who studies the structure of polymers, and mathematician David Hoffman, who studies and creates vivid computer images of both long-known and recently discovered minimal surfaces.

In one type of block copolymer, the two constituent polymers form grains consisting of stacks of alternating, equally spaced layers. Because the stacks have no preferred orientation, layers in adjacent grains may meet at any angle. When the layers happen to meet at right angles, electron-microscope images, produced from thin sections in which one of the polymers is doped to make it look darker, show a pattern resembling a comb's regularly spaced teeth.

That particular polymer structure seems to correspond to a minimal surface known as Scherk's surface, discovered in 1835 (see Color Plate 1). It can be thought of as the smooth joining of two sets of parallel planes at right angles to each other. Looking at the computed shadow, or two-dimensional projection, of that structure produces a picture closely resembling the polymer image, which itself is a two-dimensional view of the material (see Figure 1.9).

Polyisoprene and polystyrene copolymers often form interlaced networks having a tetrahedral geometry, resembling the way carbon atoms are each linked to four neighbors in a diamond structure. Thomas and his group have shown that these structures also have analogous minimal surfaces (see Figure 1.10). Such a diamondlike geometry strongly influences the copolymer's physical properties. Whereas polystyrene by itself is stiff and brittle and polyisoprene is rubbery, the combination ends up with the stiffness of one component and the toughness of the other. Such synergistic combinations may be useful in the production of superior composite materials.

Figure 1.9 Scherk's surface (*top left*), discovered in 1835, is a plausible model for representing the structure of a material consisting of two different but bonded polymers that prefer to have as little contact as possible yet are inextricably linked. In such mixed polymers, molecules aggregate into sheets that appear as parallel planes. Typically, a solid sample contains grains, each grain consisting of parallel sheets having a particular orientation. At a boundary between two grains, one would expect some sort of blending of the planes, perhaps like that seen in Scherk's surface. In fact, electron-microscope images of a polymer grain boundary show a comblike pattern (*bottom*) closely resembling the computed pattern resulting from an edge-on view of Scherk's surface (*top right*).

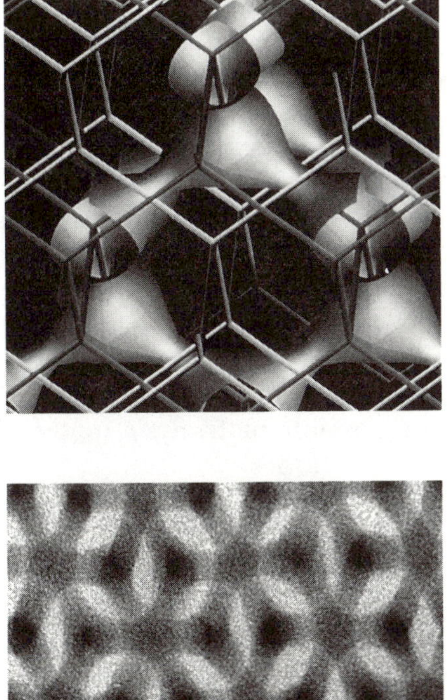

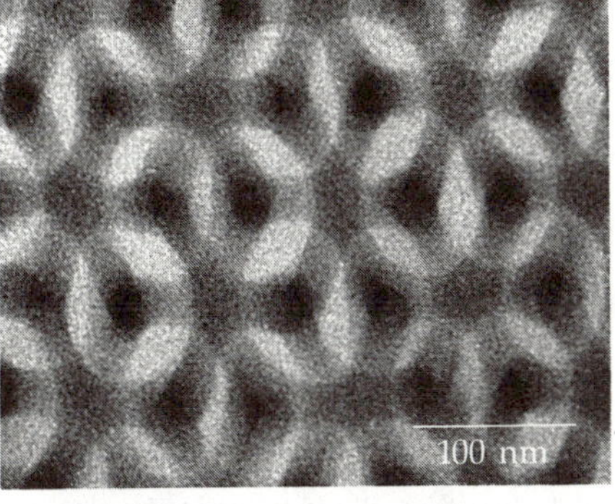

Figure 1.10 Polymer scientists have discovered a diamondlike structure, based on repeating, interlaced patterns of tetrahedrons, that serves as a good model for the interface between the constituents in a two-component polymer (*top*). Electron-microscope images of such a polymer mixture produce a pattern (*middle*) strikingly similar to that computed for the corresponding mathematical surface (*bottom*).

The researchers believe that this type of analysis would work on any physical or biological system that segregates materials. Nature is full of examples of such minimal-surface interfaces, from the surface formed between oil and water to the surface an animal or plant presents to its environment.

The research on polymer structures also suggests new mathematical questions. In a phase transition, a material changes from one form into another, which sometimes has a different symmetry or arrangement. Mathematically, the problem is one of describing the transformation of one minimal surface into another. Is it possible to take a particular structure, deform it to produce another structure, and do it smoothly and in an efficient manner? Researchers would like to know the possible pathways a material could follow to make that kind of transition.

Visualizing exotic geometrical shapes has played an important role in mathematician William Thurston's innovative attempt to classify three-dimensional surfaces, or three-manifolds. Three-manifolds can take on a bewildering array of complex shapes, and the classification of these forms has stymied many a mathematician in the past. In addition, such forms have surprising connections with other fields of mathematics, such as fractal geometry, group theory, and minimal surfaces.

Two-dimensional surfaces are easy to picture as the surfaces of three-dimensional solid objects: the surface of a sphere and the surface of a doughnut, or torus, are both two-dimensional surfaces. A three-dimensional manifold, or three-manifold, can be thought of as the three-dimensional "surface" of a mathematical object occupying four-dimensional space. One example is the space around a loosely knotted string.

Three-manifolds come in a much greater variety than ordinary surfaces. Just think of all possible knots, and then all possible knots intertwined with other knots. The imaginable complications are endless. And knots are the models for only a few of the simpler kinds of three-manifolds (see Figure 1.11).

Such a zoo of geometrical forms would seem impossible to catalog, but about a decade ago, Thurston proposed a revolutionary conjecture allowing mathematicians, in principle, to list many different types of three-manifolds and at the same time analyze

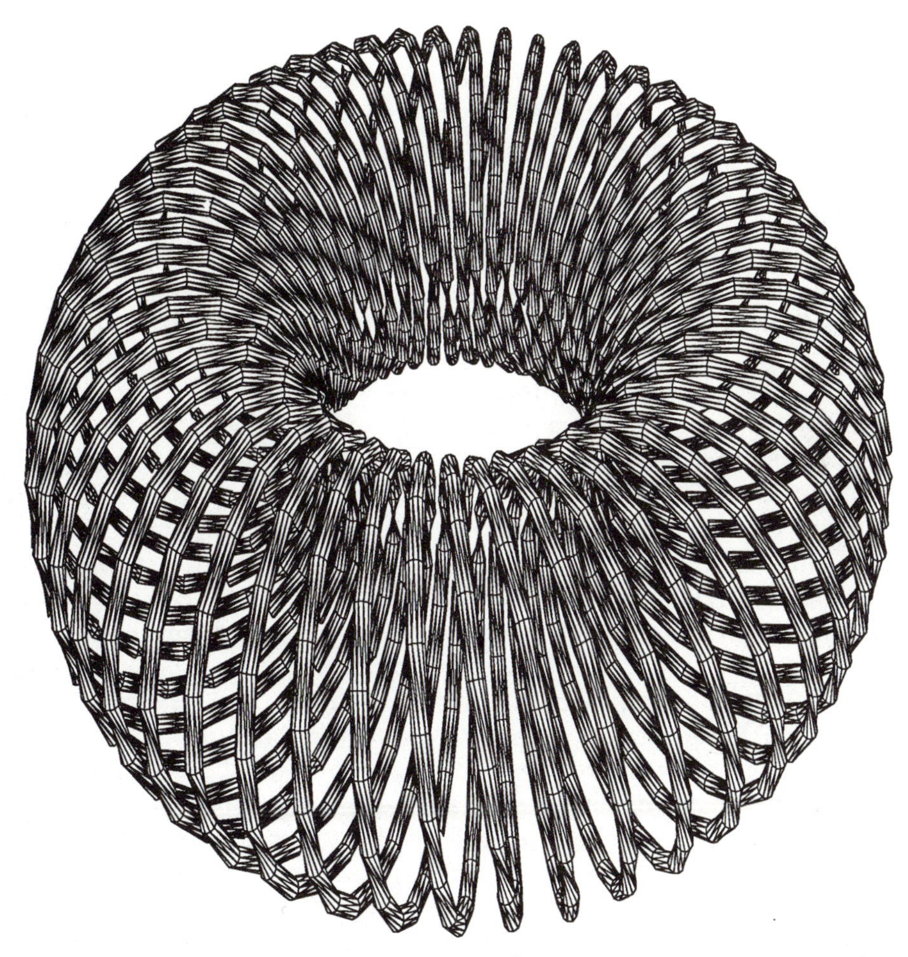

Figure 1.11 Imagine winding a piece of wire on a ring, connecting the wire's ends, then dissolving away the ring. The resulting twisted loop would resemble what is known mathematically as a torus knot. Torus knots are among the forms mathematicians study in their efforts to classify and understand three-dimensional surfaces.

their shapes. His idea was that by looking at small pieces, or neighborhoods, within a manifold, mathematicians could put each piece of the manifold into one of eight categories, based on the geometric shape of the neighborhood. It's like having eight basic sets of clothing that fit anybody in the world. The trick is to prove that the eight geometric categories suit all conceivable three-manifolds. Thurston's conjecture has been proved for wide classes of three-manifolds, and it has been tested on many other examples by hand or with the aid of a computer.

The completion of Thurston's catalog of three-manifolds requires not only deep theoretical insights into geometry itself but also the development of computer algorithms for computing and picturing various geometric forms. One question is how best to represent a three-manifold in a computer. Typically, mathematicians divide a given manifold into a set of hyperbolic tetrahedra and give the computer a description of how the tetrahedra are glued together (see Figure 1.12). That procedure works most of the time, but sometimes, in extreme cases, the tetrahedra get turned inside out. To avoid that difficulty, one possibility is to substitute two-dimensional minimal surfaces for the tetrahedral boundaries, yielding flexible, smoothed tetrahedra that can slide against each other. Such a scheme may be useful for randomly generating three-manifolds to test conjectures.

The use of computer graphics in geometry and other areas of mathematics is growing rapidly. Mathematicians are using computer-generated pictures to study fractals. Fractal patterns seem to provide an apt description of a large class of physical phenomena, from the fracturing of glass to the texture of surfaces (see Chapter 4).

Mathematicians are also using graphic images to explore the results of repeatedly evaluating algebraic expressions. They substitute a certain number into an expression, find the answer, then plug the answer back into the same expression, and so on, to see where the sequence of answers leads. This iterative process has led to the creation of colorful, intricate portraits of algebraic expressions, many of which show fractal patterns (see Chapter 7).

In 1987, a group of thirteen mathematicians and computer scientists established the "Geometry Supercomputer Project" to address mathematical problems that have languished unsolved for many years or even centuries for lack of computing power. It was the first large-scale attempt to bring together experts from a number of fields to work together on the same supercomputer to focus

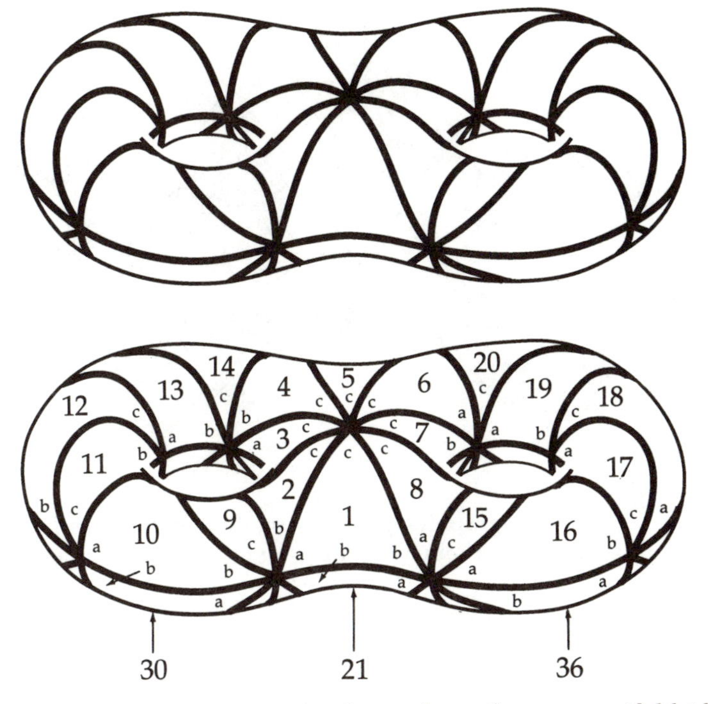

Figure 1.12 One way to represent the surface of a two-manifold (the two-dimensional surface of a three-dimensional object) like the one shown is to divide the surface into curved triangles (*top*), which are then numbered. Each triangle's three vertices are labeled a, b, and c (*bottom*). Recording in a table which edges of neighboring triangles match provides enough information to reconstruct the original surface. Three-manifolds (the three-dimensional "surfaces" of four-dimensional objects) are more difficult to visualize and handle. In this case, imagine dividing a given three-manifold into curved tetrahedrons, which are then numbered and labeled. A table specified which edges of neighboring tetrahedrons are glued together to generate that particular three-manifold. Computers work with such higher-dimensional geometric shapes simply by manipulating their corresponding tables.

on solving essentially geometrical problems, from the classification of three-dimensional surfaces, or three-manifolds, to a search for algorithms that mimic the pathways followed by nerve signals governing visual memory in humans. In the latter case, mathema-

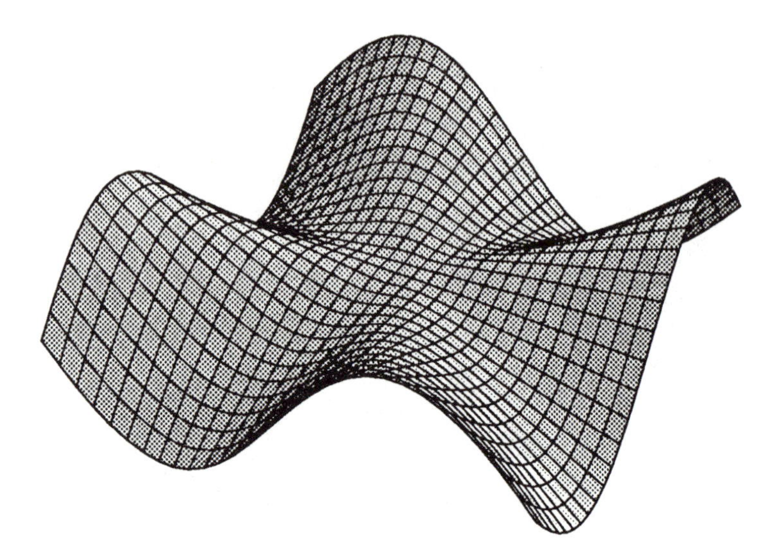

Figure 1.13 Computer graphics provides mathematicians with an important tool for visualizing mathematical forms defined by particular equations.

ticians are interested in finding out which ways of describing mathematically similar shapes would be most useful in simulating rapid and precise memory and recognition.

"I hope this project establishes for good among mathematicians the realization that the computer is an extraordinarily useful tool for exploring geometrical problems and making conjectures, and for communicating intuitions to other people," comments Benoit B. Mandelbrot, a participant in the project and founder of the study of fractals.

Because supercomputers can calculate and display in visual form the roots of equations and other mathematical objects, computer-generated pictures are also started to prove useful in areas of mathematics far from geometry (see Figure 1.13). For the first time, mathematicians are getting to "see" the content of the abstract theorems they prove and to use their eyes, not just their minds, to probe new mathematical territory and to bag new game.

A Mystery

As an intellectual endeavor, mathematics is deep rather than broad. Theorems build upon other theorems; abstract concepts are assimilated and then used as mathematical objects. Because of the extent to which new results are built on earlier results, mathematics puts a great store on truth and accuracy. If earlier results are only usually true, or almost true, the entire structure of later results is called into question and is likely to collapse. Such a deep intellectual structure is possible only because mathematics is much more self-contained than other sciences or branches of knowledge.

What surprises scientists, especially physicists, is how mathematics—seemingly a purely logical, self-contained, deductive system—corresponds so closely to the scientist's picture of nature. In many cases, the mathematical structures that arise in the laws of nature, as far as they are known, are often mathematical structures that were provided by mathematicians long before any thought of applications to physics or science arose.

Physicist Eugene P. Wigner raised the question in a famous 1960 essay called "On the Unreasonable Effectiveness of Mathematics in the Physical Sciences." Why does mathematics do so well in describing and predicting physical phenomena that no one has ever seen before? Wigner wrote: "The enormous usefulness of mathematics in the natural sciences is something bordering on the mysterious, and there is no rational explanation for it. It is not at all natural that 'laws of nature' exist, much less that man is able to discover them. The miracle of the appropriateness of the language of mathematics for the formulation of the laws of physics is a wonderful gift which we neither understand nor deserve."

Einstein's work on the general theory of relativity is one example of the remarkable connections between physics and mathematics. Einstein's recognition that gravity is merely a consequence of the curvature of space and time depended on mathematics developed by Bernhard Riemann nearly a century before (see Figure 1.14). In contrast, Riemann's work on the mathematics of curved space was motivated not by some pressing practical

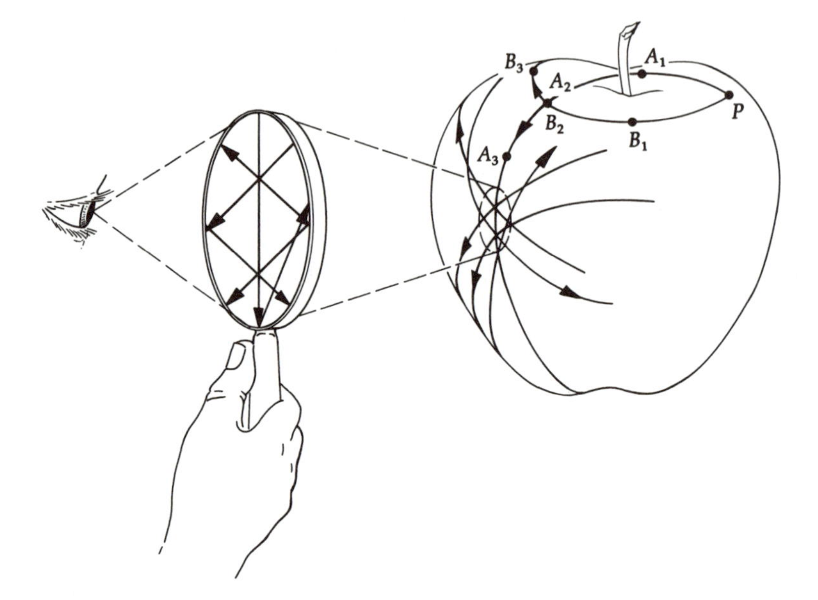

Figure 1.14 Here, the two-dimensional surface of an apple symbolizes the Riemannian geometry of the four-dimensional space-time of general relativity. For a sufficiently small region of space-time, the geometry is practically flat, as represented on the apple's surface by the straight-line course of tracks seen through the magnifying glass. In a larger region, the surface's curvature makes itself felt. According to Einstein, a particle responds to the geometry of space-time simply by following the straightest path it can. A particle's path curves because, on a global scale, space-time is curved. Newtonian theory ascribes this effect to gravitation acting at a distance from a center of attraction.

need but by the joy of doing mathematics for its own sake. Einstein's achievement, in turn, thrust Riemannian geometry into the mathematical spotlight and made it a central topic of modern geometry.

Theoretical physics has continued to adopt increasingly abstract mathematical models as the logical foundation for current theories for presenting a unified picture of the fundamental forces and particles in nature. String theory is the latest leading contender for reconciling quantum field theory with general relativity and gravitation. In this ambitious theory, unimaginably thin strings in 10 dimensions replace the dimensionless points of four-dimensional space-time. But the extra dimensions are curled up into closed spaces so small that we don't notice them and can't detect them.

Physics this fundamental is intimately intertwined with research at the frontiers of mathematics. It is in this realm that new patterns are being discovered and new edifices constructed. It's a bizarre world in which mathematicians borrow from physics and physicists borrow from mathematics. String theory, in particular, has attracted mathematical attention because the theory gives strong hints of new connections between hitherto separate parts of mathematics (see Chapter 2). Moreover, developing an understanding of the structure of string theory may ultimately involve a fundamental generalization of geometry.

Mathematics is really the science of patterns. Mathematical discovery begins with a search for patterns in data—perhaps in numbers or in scientific measurements but more often in geometric or algebraic structures. Generalization leads to abstraction, or patterns in the mind. Mathematical significance is measured by the degree to which patterns in one area link with patterns in other areas. Theories emerge as expressions of relationships among patterns—patterns of patterns.

In this way, mathematics follows its own logic, beginning with patterns suggested by something in the real world—measurement, computation, experiment, human needs or curiosity—then completing the portrait by finding all patterns that derive from the initial ones. The reason why mathematics has an uncanny way of providing just the right patterns for scientific investigation may be because mathematicians look for every possible pattern, and some of these patterns are bound to be useful ways of describing natural phenomena. If the search for patterns is what mathematics is all about, then the "unreasonable effectiveness" of mathematics may be reasonable after all.

Mathematics is presently in an exciting era of intense activity. By applying one part of their field to another, mathematicians are finding surprising connections between seemingly unrelated areas. There's a growing appreciation of mathematics as a unified structure, despite its seemingly disparate bits and pieces.

2

New Twists

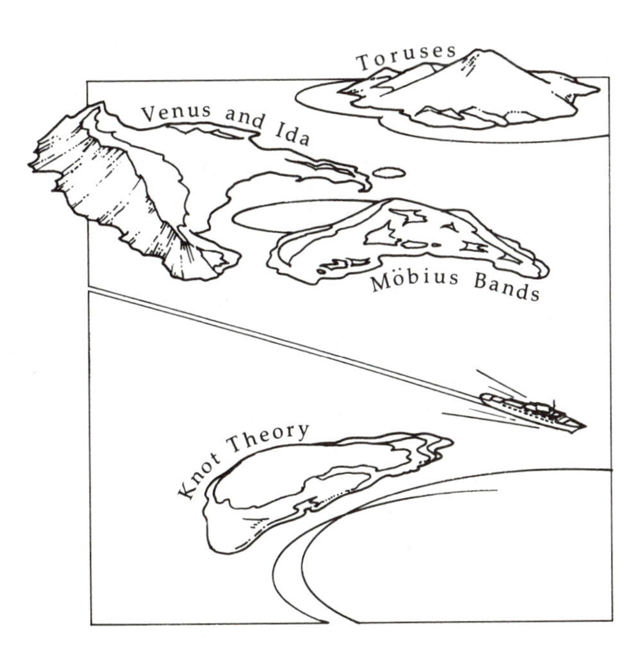

The shapes and curves of surfaces captured the imagination and attention of many mathematicians in the nineteenth century. To unfold the visual secrets compressed and hidden within the shorthand of algebraic expressions, nineteenth-century geometers drew pictures, constructed models, and even wrote manuals on how to visualize geometric shapes. Their elegant drawings and graceful plaster and wooden models (see Figure 2.1) remain a vivid testimonial to their work and to the beauty of mathematical forms. "It is the joy of form in the highest sense that makes the geometer," proclaimed nineteenth-century mathematician Rudolf Clebsch.

Such models and drawings were not only a source of pleasure but also a valuable tool for probing a slew of exotic geometric structures. The illustrations were a way of bringing together the logically abstract and the visually concrete in mathematics. They served as

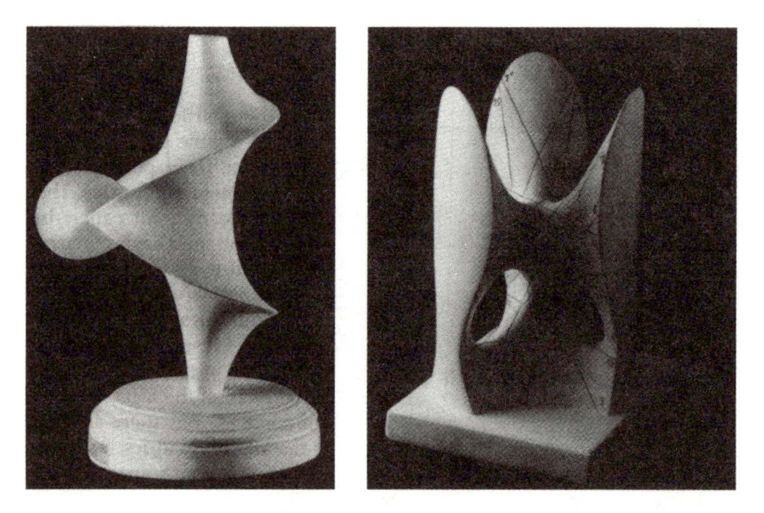

Figure 2.1 Nineteenth-century geometers often constructed wooden or plaster models of mathematical objects. Kuen's surface (*left*) features a striking blend of the angular and sinuous. Clebsch's diagonal surface (*right*) arose from the study of special algebraic equations.

landmarks in the struggle by mathematicians to understand what seemingly isolated examples were trying to tell them about the fundamental principles of geometry.

Nonetheless, by the turn of the century, pictures began to disappear from the study of geometry and to be replaced by more abstract constructs. Mathematicians, concerned that pictures sometimes led to misunderstandings and false proofs, turned to algebra, group theory, and calculus to express geometric notions. By the mid-twentieth century, mathematicians had completely forgotten how to draw or construct any but the most rudimentary figures.

Now, computers, with their remarkable ability to convert equations into colorful images on a video screen, are starting to make mathematical pictures respectable again. Pioneering work by a small group of mathematicians has demonstrated the usefulness of computer graphics for communicating ideas, discovering patterns, and suggesting new conjectures worth testing. Pictures have once again come to play a significant role in geometry, not so much for proving theorems as for illustrating them.

The Etruscan Venus

The Möbius strip, or band, is a familiar yet still intriguing object in mathematics. Discovered in 1858 by German astronomer and mathematician August Möbius, it can be constructed simply by gluing together the two ends of a strip of paper after giving one end a half twist (see Figure 2.2). The surprising result is a three-dimensional form that has only one side and one edge.

The Möbius band is one example of a wide variety of geometrical forms that play important roles in topology. Topologists emphasize the properties of shapes that remain unchanged, no matter how much the shapes are bent, stretched, twisted, or otherwise manipulated. Such transformations of ideally elastic objects

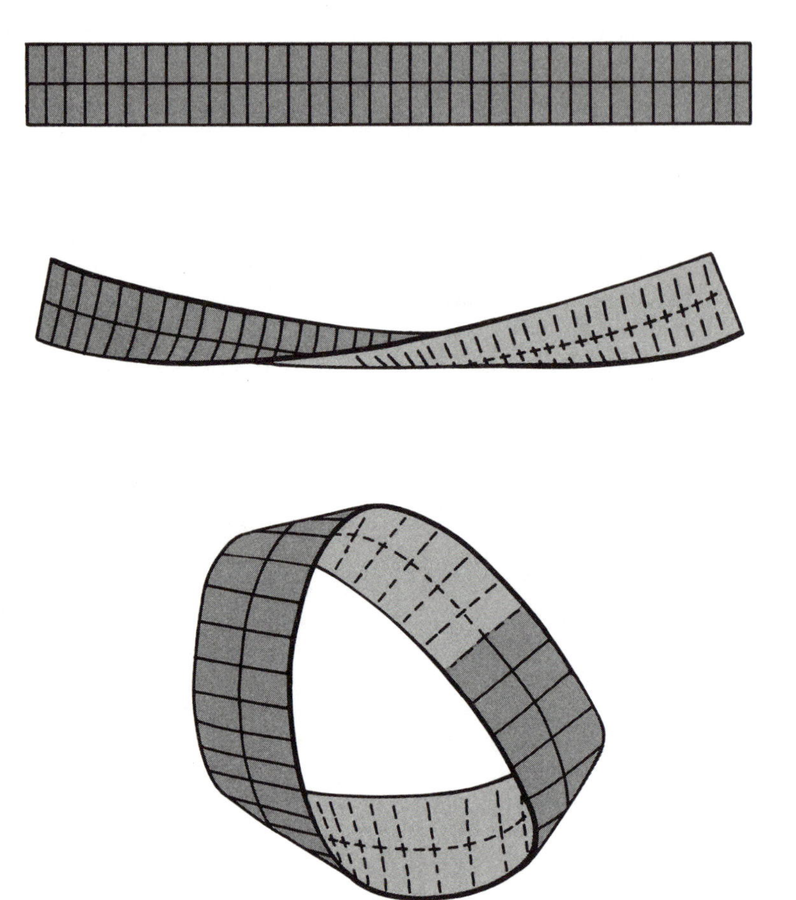

Figure 2.2 The Möbius band is the simplest one-sided, single-edged surface. It can be constructed by pasting together the two ends of a long, rectangular piece of paper after giving one end a half-twist. A bug crawling along this surface, keeping to the middle of the strip, would return to its starting position upside-down.

are subject only to the condition that nearby points on their surfaces remain close together in the transformed objects. That condition effectively outlaws transformations involving cutting and gluing. For example, a doughnut and a coffee mug are topologically equivalent. Because both forms have exactly one hole, one can imagine smoothly deforming a doughnut-shaped piece of clay, while preserving the hole, to produce a mug with a single handle. On the other hand, there is no way, short of radical surgery, to turn a spherical balloon into an inner tube.

Like a Möbius strip, a hemispherical bowl has a single edge. If a disk, also having just one edge, is sewn to a hemisphere so that the edges of each are joined, the result is a closed, two-sided surface topologically equivalent to a sphere. It's easy to imagine inflating this stitched object into the shape of a ball (see Figure 2.3, top).

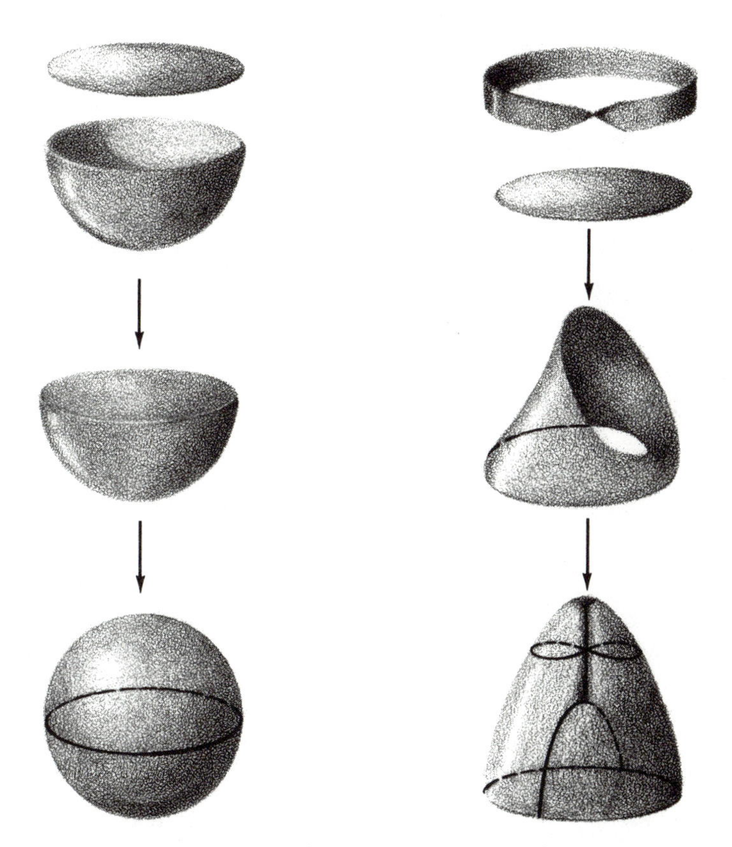

Figure 2.3 Sewing the edge of a disk to the rim of a hemispherical bowl produces a sphere (*left*). Stitching the single edge of a Möbius band to the edge of a disk twists the surface into a form known as a cross-cap (*right*).

What happens if a Möbius strip is sewn to the edge of a disk? The result is a strange, closed surface that has no true inside or outside. Pictured in three dimensions, this twisted, hybrid Möbius surface would have to cross itself somewhere (see Figure 2.3, bottom).

Topologically, the Möbius surface is a deformed version of what is known as the *projective plane*. This abstract structure can be imagined as a spherical surface in which diametrically opposite points are paired. In other words, each point of the projective plane corresponds to two points on a sphere (see Figure 2.4, top). In the projective plane, as in perspective drawing, parallel lines converge to what are termed ideal points, all of which lie on an ideal line at infinity.

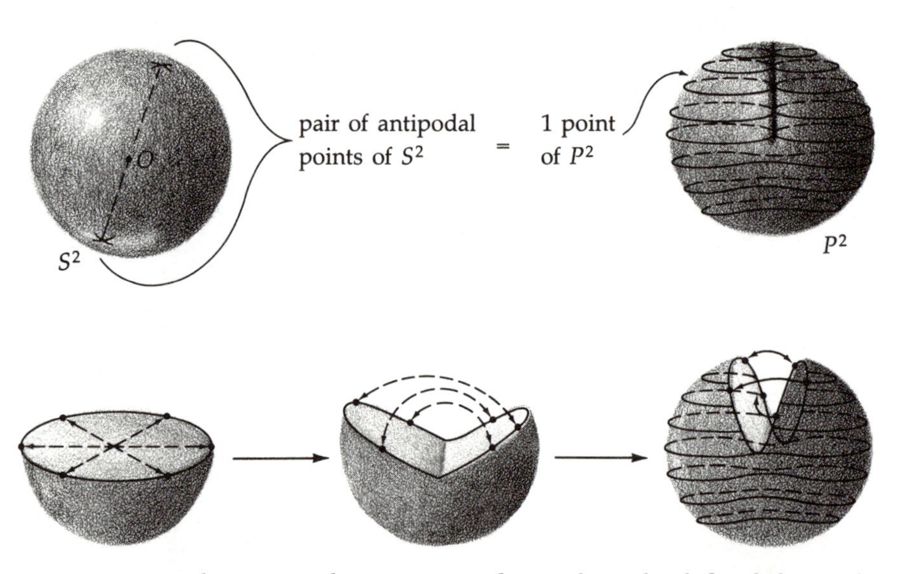

Figure 2.4 Each point on the projective plane, P^2, can be defined abstractly as a pair of points located at opposite ends on the surface of sphere, S^2 (*top*). Because each point of P^2 appears twice in S^2, it's enough to consider, say, the southern hemisphere of S^2, completing the picture by drawing together antipodal points on the equator (*bottom*). Thus, the topology of P^2 can be described as a sphere with a cross-cap (*top right*).

A curious figure known as a *cross-cap*, which appears in Figure 2.3 and looks like a dented, brimless hat, helps to provide an alternative view of the projective plane. The cross-cap results from pulling together antipodal points on a sphere's surface (see Figure 2.4, bottom). Represented in this form, the ideal line of the projective plane corresponds to the line where the cross-cap appears to penetrate itself. This line marks the pinch points where the orientation of the surface abruptly reverses.

Another possible answer to the Möbius-band-stitching problem is found among geometrical constructions made by German mathematician Jakob Steiner during the early part of the nineteenth century. His Roman surface, named in memory of a particularly productive stay in Rome, also fits the requirements of a Möbius band sewn to a disk. Steiner's Roman surface, at least from one viewpoint, looks like a severely deformed bowl (see Figure 2.5, left).

In 1900, the German mathematician Werner Boy discovered an intriguing surface that meets the same topological criteria as Steiner's Roman surface. The Boy surface looks like a strangely twisted pretzel (see Figure 2.5). However, Boy couldn't find the algebraic equations that would specify the location of every point

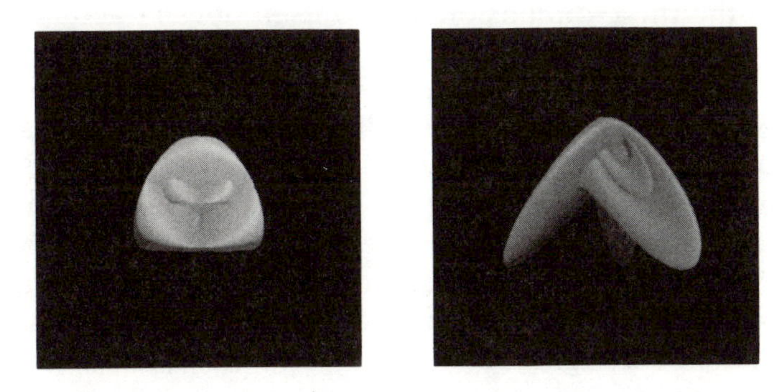

Figure 2.5 Jakob Steiner's Roman surface (*left*) is one answer to the question of what happens when a Möbius band is stitched to the rim of a hemispherical bowl and the resulting figure is suitably stretched and deformed. Werner Boy's surface (*right*) represents an alternative way of picturing what happens when a Möbius band is sewed to the edge of a hemisphere.

defining its convoluted shape. All he could do to say what the surface looked like was to draw slices of it. In the same way that slicing through a doughnut or a cone in different directions produces different forms, successive slices in different directions through Boy's surface reveal an array of strikingly different forms.

Such cross sections were sufficient for constructing a wire framework or sculpting a plaster model but not good enough for specifying a formula to describe the surface. It was not until 1984 that French mathematician François Apéry, guided by topologist Bernard Morin, finally worked out a set of equations for representing Boy's surface.

Because the Roman surface and the Boy surface are closely related (both are ways of representing the projective plane), it was natural for mathematicians to look for orderly ways of transforming one surface into the other. Such a procedure, known as a *homotopy*, involves a careful cancellation of singularities, which are places where the surface twists into itself to form a double curve or where the surface is pinched and abruptly changes direction (see Figure 2.6).

Singularities can often be removed by considering the figure as a higher-dimensional form. For example, the two-dimensional shadow of a bent wire loop sometimes appears to cross itself, although the three-dimensional loop itself doesn't actually intersect anywhere. Likewise, what appears in three dimensions to be a pinch point could very well be a perfectly regular feature in four dimensions. The pinched three-dimensional form is merely a "shadow" cast by the four-dimensional surface.

The Romboy homotopy (the name for the transformation from the Roman to the Boy surface) is based on the idea that both the Roman and Boy surfaces can be generated by moving an oval, or ellipse, through space. That's how a circle generates a sphere when rotated through 180 degrees. An ellipse, its motion governed by well-defined constraints, can stretch and contract as it sweeps out a wobbly path to produce a particular surface. The resulting bouquet of ellipses defines the shape.

Apéry, starting with the Roman surface, discovered that by smoothly altering the 10 parameters governing the angles and

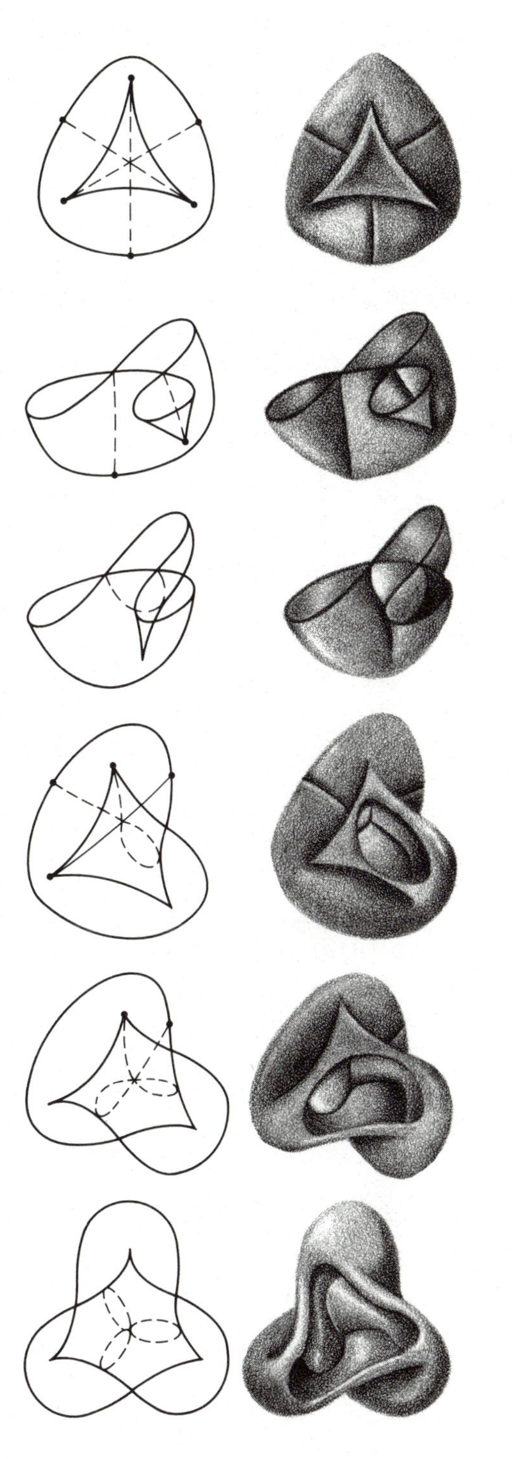

Figure 2.6 The topological transition from Steiner's Roman surface (*top*) to the Boy surface (*bottom*) is based on the cancellation of pinch points. The second and third rows in the sequence show cross sections of the Roman surface as one pinch point is canceled out. Further alterations lead to the Boy surface, sliced open to provide a clearer picture of its complex shape.

positions of an ellipse, he could gradually transform the Roman surface into the Boy surface. In fact, the equations governing this procedure show that both surfaces can be considered three-dimensional shadows of a higher-dimensional form viewed from two different vantage points.

The existence of Apéry's equations made it possible to program a computer to display the two figures and the homotopy linking them. That possibility drew mathematician George Francis into the picture. Initially, Francis programmed an Apple personal computer to generate a rough version of the homotopy. Such computer sketches are the equivalent of the wire frameworks and hand drawings frequently employed by mathematicians a century ago to study mathematical forms.

Working with artist Donna Cox and computer programmer Ray Idaszak, Francis converted his original Apple program into one that could run on a supercomputer. The researchers added shading and color to the computer drawings, putting a "skin" on what originally were little more than skeletons of the surfaces. The result was a short, 600-frame, computer-drawn film that shows a smooth deformation from the Roman to the Boy surface, then back to the original.

In the course of fiddling with the computer program, Francis also discovered a simpler way of performing the homotopy. Instead of generating the surfaces only by letting a wobbling ellipse sweep through space the whole time, Francis used a mathematical figure known as a limaçon, which looks like a closed loop coiled so that it crosses itself once to form a double loop. It can also take on a roughly heart-shaped form (see Figure 2.7).

By following the limaçon-generated homotopy, the three researchers stumbled upon a new surface that appears along the way. From one point of view, this startling apparition looks somewhat like an owl, and from another, like a female torso (see Color Plate 2). The new surface was dubbed the *Etruscan Venus* because the homotopy used is simpler or more primitive than the original Apéry homotopy and therefore stands in the same relationship as the ancient Etruscan civilization in Italy to its successor, the Roman Empire.

The Etruscan Venus turns out to be a topological form known

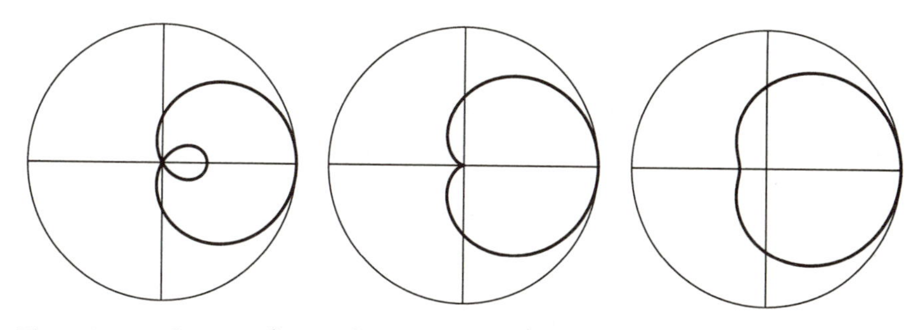

Figure 2.7 Curious figures known as the limaçons of Pascal play an important role in transforming the Roman surface to the Boy surface. In the first half of the deformation, the limaçon is a double loop, with a node at the origin of the unit circle within which the limaçon is drawn (*left*). The smaller loop finally shrinks to the cusp of a heart-shaped figure (*center*). Finally the remaining dimple (*right*) pops out to form the unit circle.

as a *Klein bottle*. It can be created by gluing together two Möbius strips along their edges, something that can readily be done in the roominess of four dimensions. In three dimensions, such a structure can be depicted only by letting the surface pass through itself (see Figure 2.8). Just as a disk is the shadow of a three-dimen-

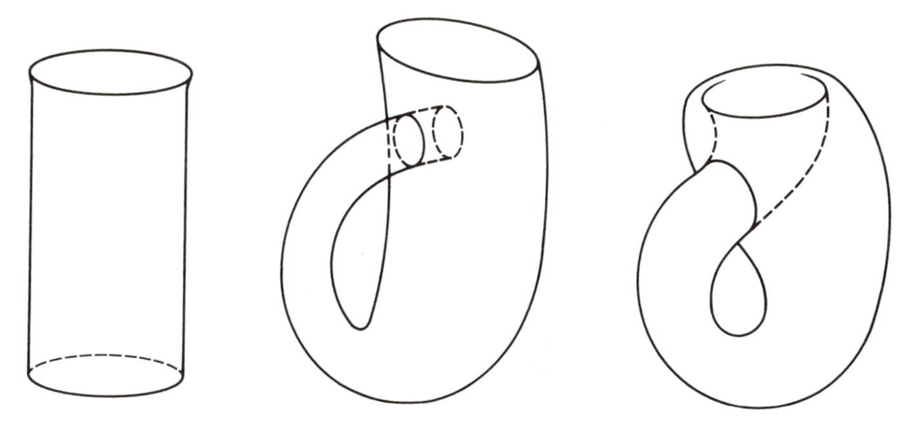

Figure 2.8 One way to create a Klein bottle is to deform a cylinder by stretching out one of its ends and poking that end through the cylinder's side before joining the two ends together.

sional sphere, the Etruscan Venus is the three-dimensional shadow of a four-dimensional Klein bottle.

Francis and his team also applied his version of the Romboy homotopy to the Boy surface generated by Apéry's homotopy. Again, they discovered a new mathematical shape. Because Idaszak was the first to see it, this one came to be called Ida. Like the Etruscan Venus, Ida is also the shadow of a four-dimensional Klein bottle (see Figure 2.9 and Color Plate 3).

In the original Apéry homotopy, both the Roman and Boy surfaces are expressions, or shadows, of the projective plane. The Francis version of the homotopy produces the Etruscan Venus and Ida, both shadows of a Klein bottle. The topological transition from the projective plane, manifested in the Roman and Boy surfaces, to the Klein bottle, as seen in the Etruscan Venus and Ida, is as severe as changing a sphere into a doughnut. That transformation, though smooth, requires the equivalent of punching out a hole in the surface and reconnecting the torn edges in a new way.

The collaboration between mathematician Francis and artist Cox proved fruitful for both. Cox gained the use of an interactive computer program, strongly rooted in topology, which could be used as a kind of sculpture machine for generating dramatic and beautiful shapes. By manipulating the 10 parameters, or dimensions, in the equations defining the Romboy homotopy, she can create an endless array of entirely new forms.

For Francis, the project generated some new mathematics and effectively demonstrated the important role that computer graphics can play in mathematical research. In his book *A Topological Picturebook*, Francis writes: "It is . . . in the making of the model, in the act of drawing a recognizable picture of it, or nowadays, of programming some interactive graphics on a microcomputer, that real spatial understanding comes about. It does this by showing how the model is generated by simpler, more familiar objects, for example, how curves generate surfaces."

Pictures without formulas mislead; formulas without pictures confuse. These statements summarize the thinking of mathematicians like George Francis, who are helping to bring the visual element back into geometry and to make pictures an essential

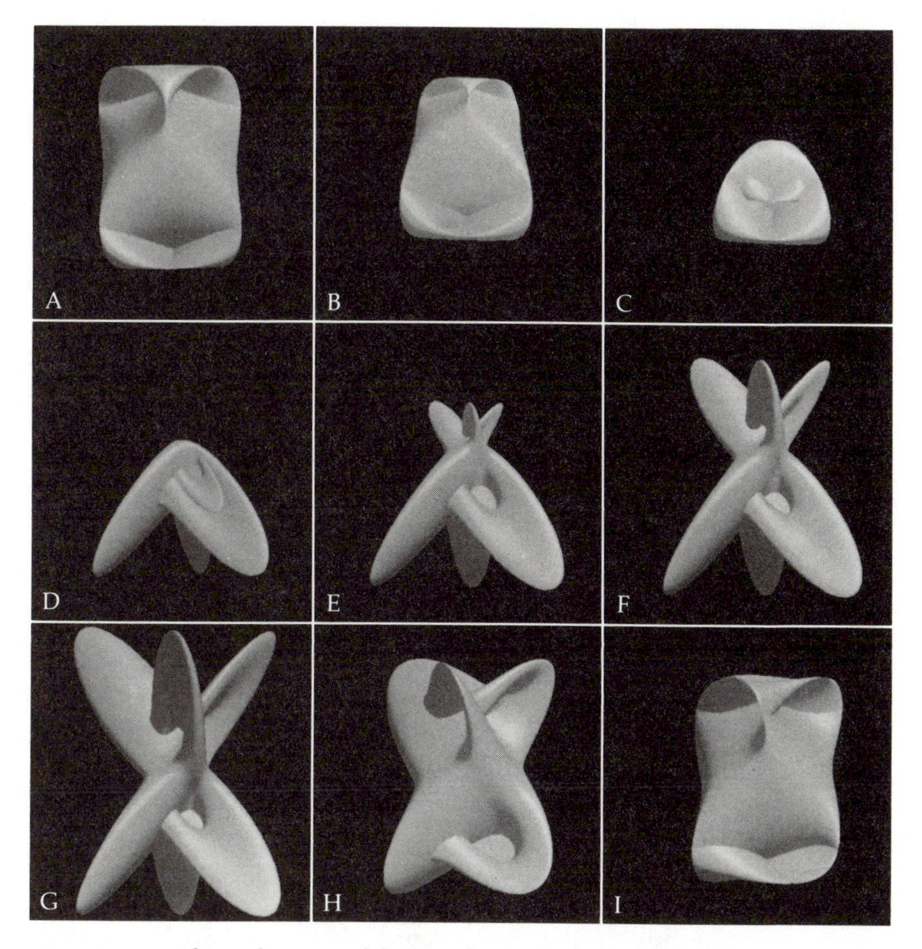

Figure 2.9 This selection of frames from the videotape "Metamorphosis: Shadows from Higher Dimensions" illustrates steps in the Romboy homotopy. The Etruscan Venus (A) is transformed first into the Roman surface (C). Next, the Roman surface is turned into the Boy surface (D), which in turn evolves into a new surface called Ida (G). Finally, Ida turns into the Etruscan Venus (I or A), completing the cycle.

part of doing mathematical research. This kind of geometric exercise is a helpful reminder that visualization is the source of all geometry—in the shapes and patterns that we see around us, in the forms we create and manipulate, and in our own sense of the space in which we live.

Inside Out

Turning an inflated beach ball inside out means letting out the air, reversing the ball's surface by pulling it through the opening, pumping in more air, and finally resealing the ball. But a mathematical sphere has no air valve or orifice. Turning such a sphere inside out, without tearing or cutting it open, seems intuitively impossible.

Mathematicians, however, can play such mental games by their own rules. If it were possible to push a surface through itself, meaning that two points on the surface could temporarily occupy the same point in space, then a solution to the problem of reversing, or everting, a sphere's surface might exist. The trick is to perform the entire transformation, or eversion, without letting a crease enter the picture.

Intuition suggests that the mathematical problem of everting a sphere without allowing creases can't be solved. Imagine a sphere painted blue on the outside and red on the inside. Pushing the north and south poles toward the sphere's center, then past each other, forces the original inner surface to protrude more and more. The transformed object begins to look like a red sphere with a blue tube running around its equator. Gradually the tube (the remaining portion of the outside) becomes thinner and thinner until it vanishes. That leaves a sphere with a red outside and a blue inside. Unfortunately, during the transformation, the tube forms a tight loop that must be pulled through itself. That produces a sharp crease, which isn't allowed (see Figure 2.10).

Can an eversion be done by a continuous, smooth process without having to introduce any creases or pinch points over the

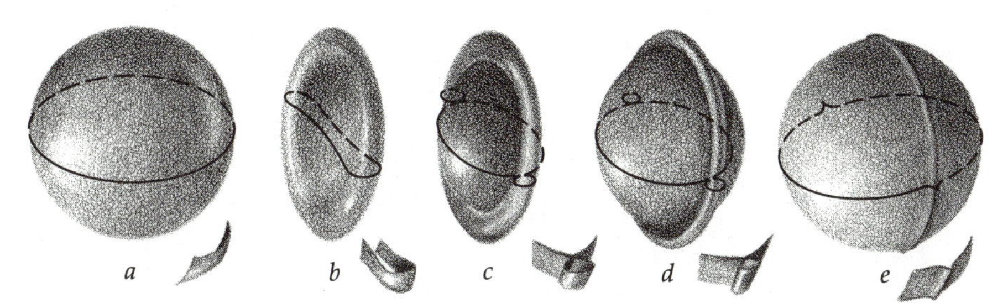

Figure 2.10 An inadmissible procedure for turning a sphere inside out entails pushing regions on opposite sides toward the center (b) and then through each other. The original interior begins to protrude on two sides (c); these two sides are pulled out to form a sphere (d and e). Unfortunately, when the looped portion of the original surface is pulled through itself, a crease is introduced in the surface, and that's not allowed.

course of the transformation? Intuition again suggests that the problem can't be solved, and mathematicians struggled for nearly a century to resolve this famous question one way or the other. That a problem concerning an object as simple as a sphere seemed so difficult to settle was both a mystery and a challenge to many mathematicians.

The first part of the answer finally arrived in 1959, when mathematician Stephen Smale, then a graduate student, proved an abstract theorem that indirectly leads to the proposition that sphere eversions are possible. The result was so surprising that even Smale's thesis adviser was skeptical and insisted there had to be a mistake. But the logic of Smale's proof held up.

Smale's effort demonstrated it was possible, in principle, to piece together the myriad, minute geometrical constructions specified in his proof to assemble an explicit picture of how to turn a sphere inside out without introducing a crease anywhere along the way. But Smale's step-by-step path for accomplishing a sphere eversion was so complicated that no one could visualize his procedure. Thus, for some time after Smale's discovery, mathematicians knew that turning a sphere inside out was possible, yet no one had the slightest idea how to do it.

Gradually, visual answers to the sphere-eversion problem began to emerge. One of the first practical eversion schemes, proposed by Arnold Shapiro, starts with the pushing together of two regions on opposite sides of the sphere toward the center and then through each other so the interior now protrudes. The surface is then stretched, pinched, and twisted through several highly complicated intermediate stages before the eversion is complete. Shapiro himself never drew pictures of the crucial stages in his transformation but relied on topological arguments to make his case.

Eventually, a number of mathematicians figured out workable schemes for visualizing a sphere eversion. These schemes involved complex sets of moves that carry the surface through the key steps in the transformation (see Figure 2.11). And mathematicians continued to look for simpler means of describing and displaying how the change occurs.

In the 1970s, the task of developing a more straightforward way of doing a sphere eversion fell to French topologist Bernard Morin. In his history of sphere eversions, mathematician George Francis writes: "Bernard Morin is not distracted, like the rest of us, by pencil and paper and the business of drawing and looking at pictures. He is blind. With superb spatial imagination, he assembles complicated homotopies of surfaces directly in space. He keeps track of temporal changes in the double curves and the surface patches spanning them. His instructions to the artist consist of a vivid description of the model in his mind."

Morin succeeded in developing and extending Shapiro's approach, putting together what can be thought of as a set of architectural plans for a sequence of three-dimensional constructions showing the essential steps in a sphere eversion. In 1977, Nelson Max, with the aid of spectacular computer graphics, used the coordinates from meticulously soldered wire models of the various stages in an eversion to produce a dramatic animated film vividly illustrating the transformation.

But neither Morin nor Max provided mathematical formulas describing the full transformation. Mathematicians could look at the pictures but they couldn't get at the mathematics of the transformation itself. The problem, still not solved, is one of finding a set of equations to describe the smooth transition that takes place

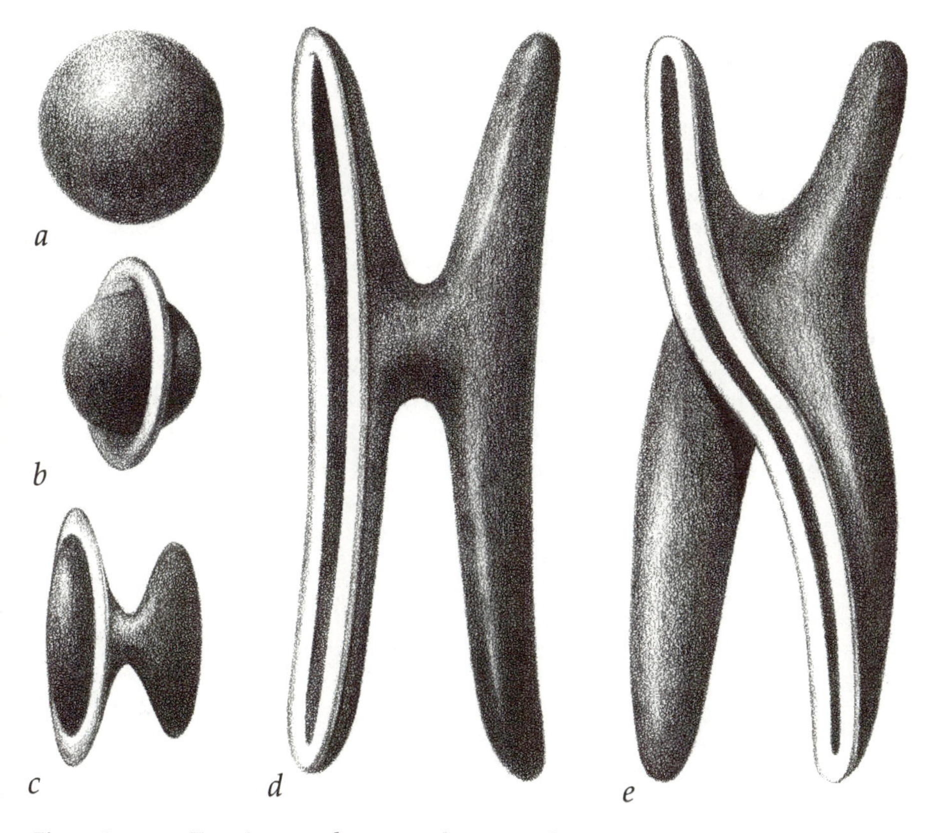

Figure 2.11 Turning a sphere inside out without introducing a crease is a complicated procedure that requires deforming the sphere in strange ways. In mathematician Anthony Phillips' version, the process starts simply enough with the pushing of opposite sides toward the center and through each other so that the interior protrudes in two regions (*b*), as it did in Figure 2.10. One part of the original interior is then distended (*c*) to give a surface resembling a saddle on two legs (*d*). The two legs are then twisted counterclockwise to give surface *e*. From that point on, the contortions get very elaborate, but in the end, the result is a sphere turned inside out, with a crease nowhere along the way.

during a sphere eversion, just to be sure that no cusps, folds, or creases lurk in the steps between successive pictures.

In 1989, Morin discovered what may be the simplest possible route for a sphere eversion. Using Morin's method, one need follow the positions of only 12 points on a sphere. This means that for the purposes of this problem, a sphere is equivalent to a polyhedron with 12 corners, or vertices. The coordinates of those points throughout the transformation provide all the information a mathematician needs to understand the eversion.

Morin starts with a cuboctahedron, which looks like a cube with its corners lopped off (see Figure 2.12, left). This polyhedron has 12 vertices and 14 faces (6 squares and 8 equilateral triangles). By using a sequence of elementary moves (moving a vertex along

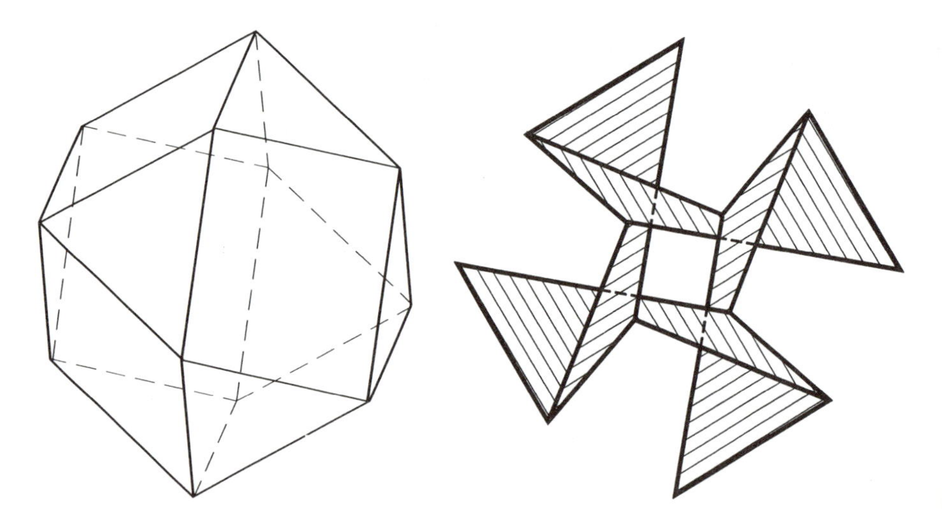

Figure 2.12 In his 12-vertex sphere eversion, Morin begins with a cuboctahedron (*left*), a polyhedron having six square and eight triangular faces. Step by step, he converts the cuboctahedron into what he calls the central model, a polyhedron consisting of four nonconvex pentagons and eight triangles (*right*). Only the ring of pentagons appears in the diagram. Manipulating the central model turns it inside out, and another flurry of moves brings back the cuboctahedron, now turned inside out.

an edge), Morin transforms the cuboctahedron into a curiously shaped figure, which he calls the "central model," with only 12 faces but the same number of vertices as before. Four of its faces are nonconvex pentagons, which look like notched quadrilaterals (Morin calls them "horses"), and the rest are triangles of two different types (see Figure 2.12, right, and Color Plate 4). A sequence of six elementary moves carries the central model through the tricky stages of the eversion. A final flurry of moves produces a cuboctahedron again, now turned inside out.

Morin's polyhedral model of a sphere eversion makes it easier to keep track of where all the pieces of a surface are going during a transformation. Focusing on just 12 vertices forces mathematicians to see the transformation's crucial twists and turns. They can ignore extraneous complications while concentrating on the eversion's essential parts.

Although Morin's version is the simplest possible polyhedral model of a sphere eversion, it isn't the first. In 1988, John Hughes created a polyhedral model having several thousand vertices as part of an effort to find a set of equations that he could use to program a computer to perform and display a complete sphere eversion.

Hughes' approach to finding an explicit formula was to build up a given surface from small patches already defined by specific mathematical expressions. It's a process akin to sewing together a quilt from miscellaneous pieces of material. Mathematically, the idea is to piece together algebraic expressions known as polynomial functions, each defining a small piece of surface, add them together, and make sure the patches meet smoothly. If you imagine trying to cover a sphere using square pieces of fabric, you get a sense of how tricky and important the final smoothing operation can be.

Using such mathematical quilting, Hughes developed equations, which he used to produce a dramatic animated film that vividly demonstrates a sphere eversion (see Color Plate 5). Such a formulation now lets mathematicians track particular characteristics of the transformation. For example, at various times during a sphere eversion, there are a number of "double" points where two surfaces intersect. Taking the family of all those double points over the course of an eversion produces a stack of curves, which when glued together is itself a surface. Mathematicians can now investigate the topology of that new surface.

Because it's an attempt to visualize a fairly abstract mathematical idea, the sphere-eversion problem is a never-ending quest. There's always something left to be done to make the pictures more clear and more meaningful.

"It's a problem that captures the imagination," says Anthony Phillips, one of the first mathematicians to work out a way of visualizing a sphere eversion. "It's a fairly easy problem to explain. It's something that somehow should be very simple to do but is really very complicated."

Turning a sphere inside out is only one example of the transformations shown to be possible by Smale's theorem. His ideas can be applied to many objects, including the torus, or surface of a doughnut, and other familiar geometric forms (see Figure 2.13). But, as in the case of the sphere eversion, knowing that something can be done isn't the same as knowing how to do it.

For all their complexity, pictures aren't the whole mathematical story. What's fascinating about Smale's published proof is that it contains no pictures at all. All the intricacies of a sphere eversion and many other transformations are expressed in an abstract language far removed from the visual world. It's a startling example of how much information mathematicians working on geometric problems can convey to one another without relying on pictures.

Knot Physics

Consider the plight of a gardener struggling with a recalcitrant tangle of garden hose. Sometimes, no amount of pulling or twisting unsnarls the coils. At other times, the tangles readily come apart, and the hose emerges unknotted. The question arises: Is

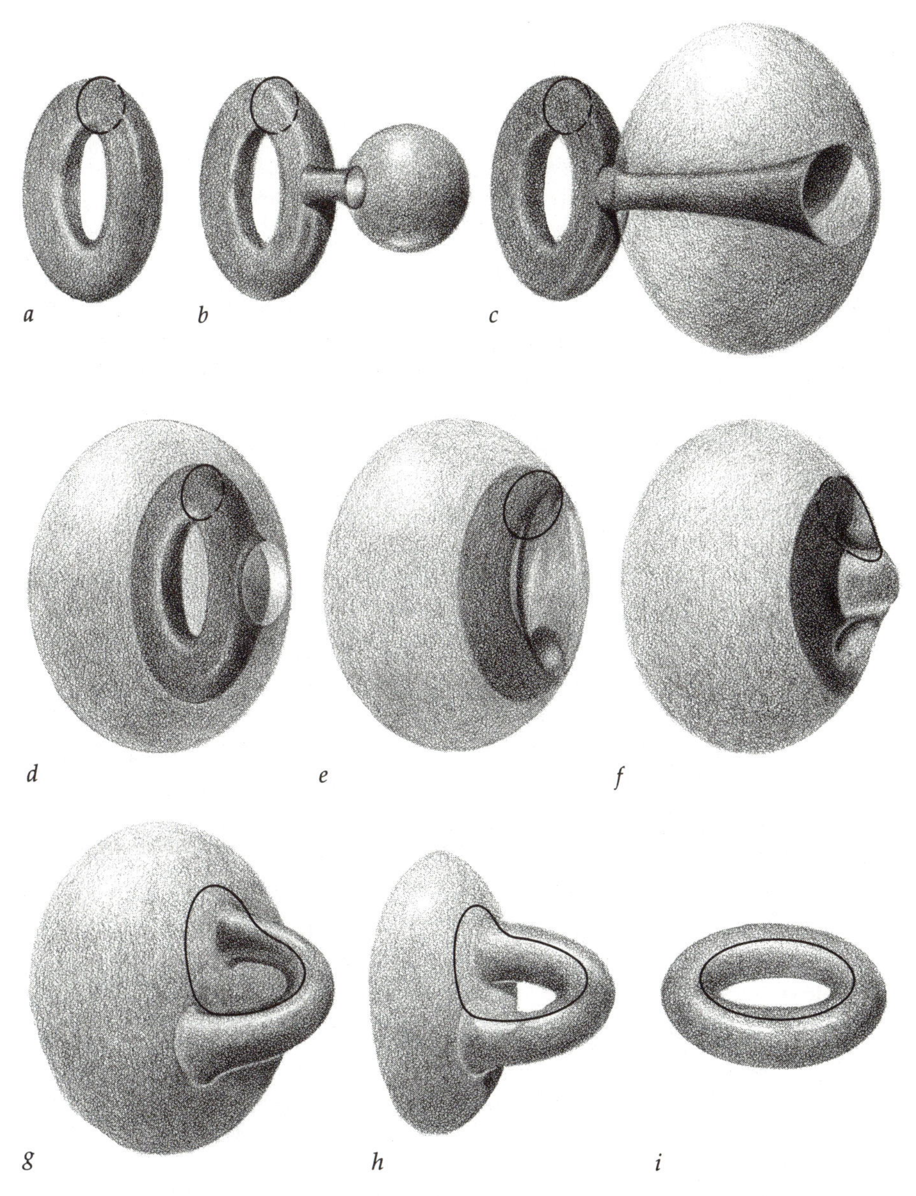

Figure 2.13 Turning a torus inside out includes the deformation required for everting a sphere. A small sphere is extruded from the torus (*b*) and everted (*c*). Then the inside-out sphere is enlarged until it engulfs the torus (*d*). Next, the tube leading from the outside of the everted sphere to the inside of the torus is enlarged and the torus pulled through it (*e*, *f*, *g*). Finally the sphere is shrunk (*h*, *i*).

there an easy way to tell the difference between an apparent tangle and a true knot?

For more than a century, mathematicians have been looking for systematic ways of distinguishing real knots from messy tangles that come apart when shaken. That problem is one of the central questions in knot theory. Knot theorists are also interested in ways of classifying all possible knots and in schemes for deciding when two seemingly different knots are really equivalent. Although a competent boy scout or sailor can readily identify a reef knot and a granny knot, mathematicians have a tougher task because they must deal with all conceivable knots. In many cases, two knots may look the same when, in fact, they are completely different. Alternatively, a knot may be so contorted that its true identity is masked (see Figure 2.14).

One apparently straightforward way to tell whether two seemingly different knots are really the same is to try twisting and pulling one knot until it matches the other. If that can be done without having to cut either knot, the two knots must be equivalent. However, failure to turn up a match, even after hours of fruitless labor, doesn't prove the two knots are different. Possibly, the right combination of moves was somehow overlooked.

Mathematically, the answer is to find a simple way to pin a label on a given knot so that either two knots with the same label are equivalent, or two knots with different labels are truly different. In the latter case, the label would be enough to indicate that no amount of twisting, pulling, or pushing would ever transform one knot into the other. An easily computed label would allow knot theorists to tell knots apart without having to go through the messier task of tangling with the knots themselves.

The last few years have seen a resurgence of interest in knot theory, precipitated by the unexpected discovery of several new ways of mathematically distinguishing knots. Mathematicians are beginning to catch glimpses of what these new methods mean geometrically. A few investigators are exploring mysterious hints of possible links between knot theory and theoretical physics.

The connection between knot theory and the physical sciences goes back to the late nineteenth century, when British scientist Lord Kelvin hypothesized that atoms were knotted vor-

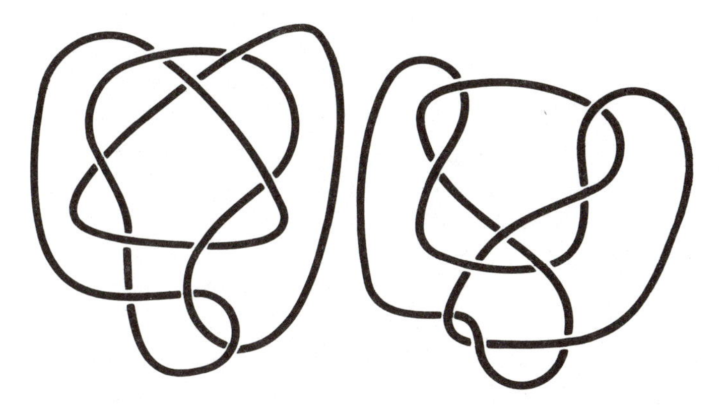

Figure 2.14 A mathematical knot consists of a single, twisted loop. The central problem in knot theory is distinguishing different types of knots, a task often fraught with difficulties. For 75 years, these two knots were thought to represent distinct knot types. Then, in 1974, it was discovered that a simple change in point of view shows the two knots are actually the same.

tices in the ether, an invisible fluid thought to fill all space. By classifying knots, he hoped to organize the known chemical elements into a periodic table.

The mathematician's idea of a knot differs somewhat from that of a sailor or a boy scout. One way to picture a mathematical knot is to think of a tangled piece of string with its two ends spliced together (see Color Plate 6). In other words, unlike a knotted piece of rope, a mathematical knot has no free ends. Otherwise, any knot could easily be untied simply by fiddling with a free end in just the right way.

To make the task of classifying knots more manageable, investigators choose to examine the two-dimensional shadows cast by the three-dimensional knots themselves. Even the most tangled configuration can be pictured as a continuous loop whose shadow winds across a flat surface, sometimes crossing over and sometimes crossing under itself. In drawings of mathematical knots, tiny breaks in the lines signify underpasses or overpasses.

One convenient measure of a knot's complexity is the minimum number of crossings that show up after looking at all possible shadows of a particular knot. A loop without any twists or crossings (in its simplest form, a circle) is called an unknot. The simplest possible knot is the overhand, or trefoil, knot, which is really just a circle that winds through itself. In its plainest form, this knot has three crossings. It also comes in two forms: left-handed and right-handed configurations, which are mirror images of each other.

By listing knots according to crossing number (see Figure 2.15), mathematicians initially hoped to identify important trends and patterns. However, that task proved to be complicated, tedious, and less rewarding than mathematicians had wished. Nevertheless, these comprehensive lists, so painfully assembled, had an enormous impact on knot theory. They provided the raw material for numerous conjectures and speculations. Because exam-

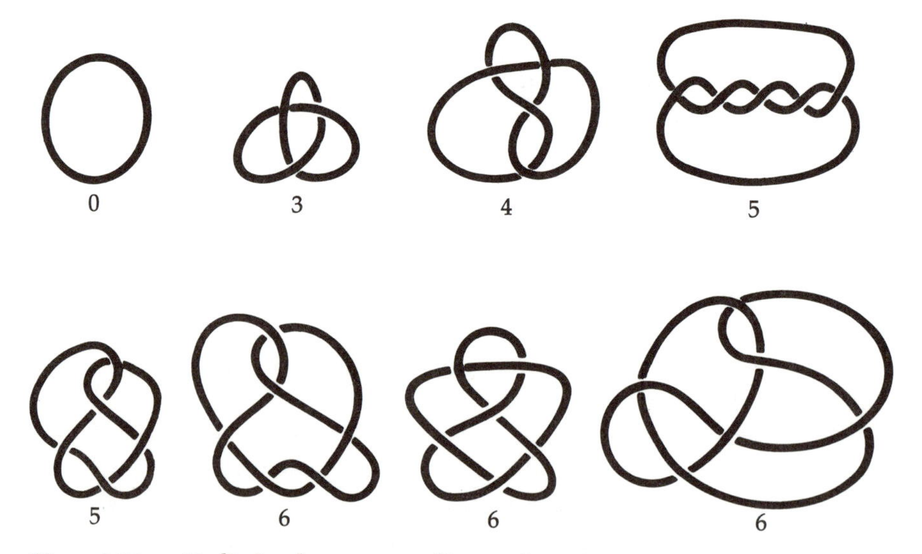

Figure 2.15 By listing knots according to the minimum number of times in which a given loop crosses itself in a two-dimensional picture of the knot, mathematicians looked for trends and patterns to help them distinguish knots. The unknot (*top left*) has no crossings, while the trefoil knot next to it has three crossings. Mathematicians have identified 12,965 distinct knots with 13 or fewer crossings.

ples were readily available, mathematicians could test their ideas and study a wide range of knot properties in depth.

One approach to labeling knots is to use the arrangement of crossings in a knot diagram to produce an algebraic expression for that knot. Such a label, which stays the same no matter how much a given knot may be deformed or twisted, is known as an *invariant*. Over the years, mathematicians developed a variety of such invariants, often expressed in the form of a polynomial algebraic expression. But the methods weren't foolproof. It was always possible that two knots with the same polynomial invariant could still be tied differently.

In 1984, Vaughan F. R. Jones unexpectedly discovered a connection between von Neumann algebras (mathematical techniques that play a role in quantum mechanics) and braid theory. A mathematical braid can be thought of as a set of hanging strings that have been twisted together in some pattern. The top and bottom ends of such a braid pattern can be tied together to form a knot (see Figure 2.16). The chain of reasoning from physics to braids to knots led Jones to the formulation of a new polynomial invariant. Jones's discovery prompted a great deal of excitement in the mathematical community because his polynomial detects

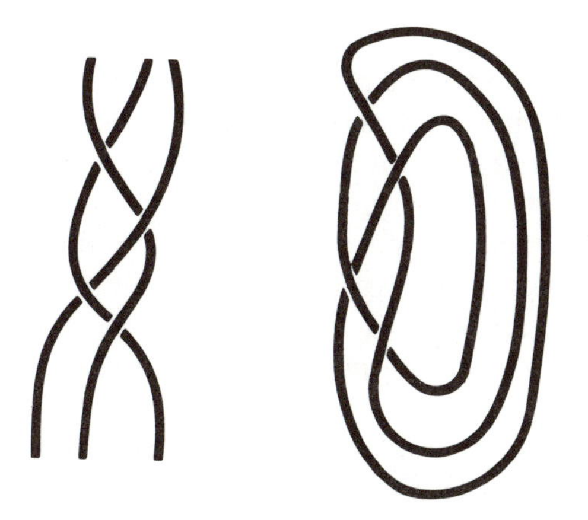

Figure 2.16 Splicing the top of a braid to its bottom forms either a single knot, as in this case, or several linked knots.

the difference between a knot and its mirror image, something that previously known invariants failed to do.

To the astonishment of knot theorists, the discovery of the Jones invariant stimulated the formulation of many more, new invariants. The invariants appear to have certain properties that tie them closely together, but those resemblances may be misleading. Nonetheless, it already seems clear that all these invariants are part of a still larger picture that mathematicians yet barely glimpse.

Although Kelvin's formulation of atomic theory in terms of knots never panned out, researchers are now seeing hints of new links between knot theory and physics. Possible connections between physics and knot theory aren't entirely surprising. Knots are physical to begin with, and knot diagrams sometimes resemble diagrams physicists use to represent interactions between elementary particles. By putting the mathematics describing the interaction of two particles at various knot crossings, a knot diagram becomes a way of summing up all the different kinds of particle interactions that can take place.

Knot invariants could also be used in statistical mechanics to analyze the molecular behavior of a condensing vapor or the lining up of electronic spins when a material becomes magnetized. Physical situations often require the summing of a lot of different interactions, and the knot invariants seem to be special kinds of averages taken over a range of different possibilities.

To find even more knot invariants, some mathematicians have also been using a set of equations known as the Yang–Baxter equations, which are sometimes used by physicists in statistical mechanics. What strikes many mathematicians as extremely mysterious is the fact that a purely mathematical device that physicists use to solve their models in statistical mechanics is just the mechanism for generating knot invariants. It's very tempting to try to find out if this mathematical relationship has any physical meaning.

If any connections exist beyond mere coincidence, they are likely to be subtle. The most intriguing possibility is a link between field theories in physics and knot invariants in mathematics. If that were the case, it would help a physicist determine whether two apparently different field theories are truly different. To evaluate theories describing, for example, the forces between certain particles, a physicist could convert the problem into a question about knots and knot invariants.

Mathematicians at Oxford University in particular have been exploring connections between the topology of surfaces and certain complicated equations used by physicists to describe the behavior of fundamental particles. In the course of this exploration, one of the researchers, Michael Atiyah, noticed a strong resemblance between Jones's knot polynomials and some aspects of quantum field theory. He suggested that the two are somehow related. Theoretical physicist Edward Witten, one of the chief proponents of string theory, took Atiyah's provocative suggestion seriously and developed a grand scheme, which appears to tie together a wide range of mathematical ideas and may lead to a deeper insight into string theory.

According to string theory, which attempts to provide a unified picture of gravity and quantum mechanics, the world is made up not of elementary particles but of tiny strings that wriggle about. Different particles, whether electrons, quarks, or neutrinos, correspond to strings vibrating in particular ways. This idea of a vibrating loop of string sitting in physical space resembles the mathematical notion of a knot as a twisted circle embedded in three-dimensional space.

Moreover, just as Einstein based his theory of gravity on the geometrical principle of the curvature of space, Witten is seeking a similar geometrical foundation for string theory. He contends that string theory will ultimately flourish as a new branch of geometry.

Witten's "topological quantum field theories" hint at what such a geometrical foundation would look like. He starts with an appropriate mathematical description of an abstract geometrical space. To that construct he applies techniques that physicists often use to introduce quantum rules into a system. Because Witten's theory has no physical content, it says nothing about particles and forces in the real world, but the results provide information about the geometric space that acts as the setting for the calculations and puts its characteristic stamp on the results.

At this point, the impact of Witten's approach on physics is still not clear. Its main potential application is in the development of a consistent theory incorporating quantum gravity. But physicists need a better understanding of both string theory and topological field theories to see exactly what the relationship is.

The impact of Witten's ideas on mathematics, however, is already evident. Because Witten's method can be applied to any twisted or deformed three-dimensional space containing a knot, his theory supplies a way of defining or identifying a knot in three dimensions. It also provides for the first time a picture of the kind of geometric information encoded in a Jones polynomial. In the past, knot theorists could calculate a Jones polynomial for a two-dimensional representation of a knot but had no idea how that information was related to the knot's three-dimensional form. The new findings may shed light on the perplexing question of why the new polynomial invariants work so well in the first place.

Mathematicians also see the possibility of a network of links joining quantum field theory to a variety of mathematical ideas, from knots and the geometry of curved spaces to algebra and group theory. For example, a few mathematicians are working on reconstructing knot theory from an algebraic viewpoint. The underlying structures are so similar that the temptation to look for a deeper connection is irresistible. Many superficially different branches of mathematics are becoming understood as different aspects of the same thing.

Mathematician John W. Morgan says, "The physicists are teaching us something that we didn't see in pure mathematics. It's having a lot of influence on very active areas in mathematics." Not surprisingly, an increasing number of mathematicians are rediscovering physics, plunging into the intricacies of quantum field theory and seeking to understand the special mathematical methods physicists have developed for solving theoretical problems related to the nature of mater and the structure of the universe.

Ultimately, the goal of the mathematical study of objects such as knots is to learn new problem-solving techniques. Such techniques often turn out to be useful even if, along the way, mathematicians labor over seemingly frivolous concerns to develop the methods and insights needed. It often happens that the solution to a specific problem is much less important than the creation of a new technique for reasoning.

A Different Dimension

Mathematicians are normally cautious and meticulous individuals, but when they describe the results of their studies of four-dimensional space, they are apt to apply adjectives like "bizarre," "strange," "weird," and "mysterious." By combining ideas borrowed from theoretical physics with abstract notions from the field of topology, mathematicians are discovering that four-dimensional space—a realm just a short step beyond our own familiar, three-dimensional world—has mathematical properties quite unlike those characterizing space in any other dimension. This type of abstract result is practically impossible to visualize. It follows logically from mathematical notions of dimension and space, part of a mental game mathematicians play in their search for patterns and relationships among geometrical structures not just in one, two, and three dimensions, but in higher dimensions as well.

In general, the term "dimension" signifies an independent parameter, or coordinate. A space has three dimensions if each of its points is completely determined by three independent numbers. For instance, it takes three coordinates, representing longitude, latitude, and altitude to specify the location in three-dimensional space of an airplane above the Earth's surface. Similarly, a space has seven dimensions if seven numbers are needed to locate a point in that space (see Figure 2.17).

The dimensions themselves have no intrinsic meaning. Physicists often choose a set of dimensions in which the first three represent independent directions in physical space and the fourth is time, but that's only one of innumerable possibilities. The four

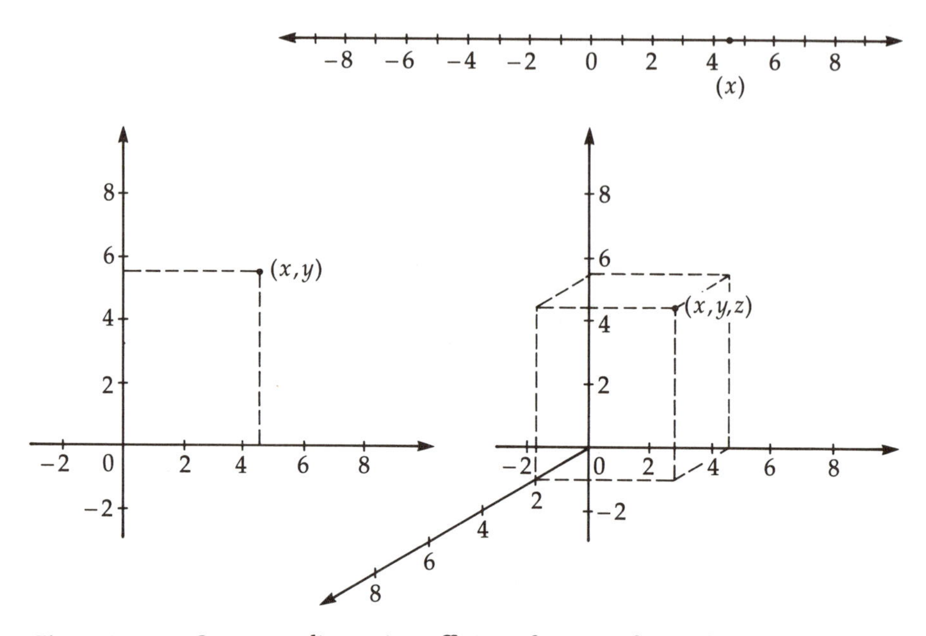

Figure 2.17 One coordinate is sufficient for specifying the location of a point along a one-dimensional line or curve; two coordinates suffice for specifying a point's location on a two-dimensional surface; and three coordinates suffice for specifying a point's location in three-dimensional space.

dimensions could just as well be pressure, volume, temperature, and mass, or any other set of four parameters.

The simplest mathematical spaces are known as euclidean spaces. An infinitely long line is a one-dimensional euclidean space. The plane is two-dimensional. We think of the space in which we live as three-dimensional.

The term *manifold* covers somewhat more complicated types than euclidean spaces. In everyday parlance, a manifold is a pipe or chamber bristling with subsidiary tubes. Its mathematical manifestation encompasses surfaces that logically appear flat, or euclidean, but on a larger scale may bend and twist into exotic and intricate forms.

The Earth's surface resembles a two-dimensional mathematical manifold, or two-manifold. An inhabitant of the Great Plains sees essentially a flat surface, whereas an astronaut orbiting the Earth sees the rounded surface of a sphere. Any surface, however, curved and complicated, so long as it doesn't intersect itself, can be thought of as consisting of small, two-dimensional, euclidean patches glued together (see Figure 2.18).

Manifolds occur throughout most of mathematics. For example, they arise as the solutions of simple equations such as $x^2 + y^2 + z^2 = 1$, which describes a sphere. They also play an important role in physics. Higher-dimensional spaces are often a convenient vehicle for expressing important aspects of the physical universe.

One example of such a physical manifold is the space of all possible locations of three balls on a billiard table. Because it takes two numbers to specify the position of each ball, six numbers are needed to specify the positions of all three balls. Hence,

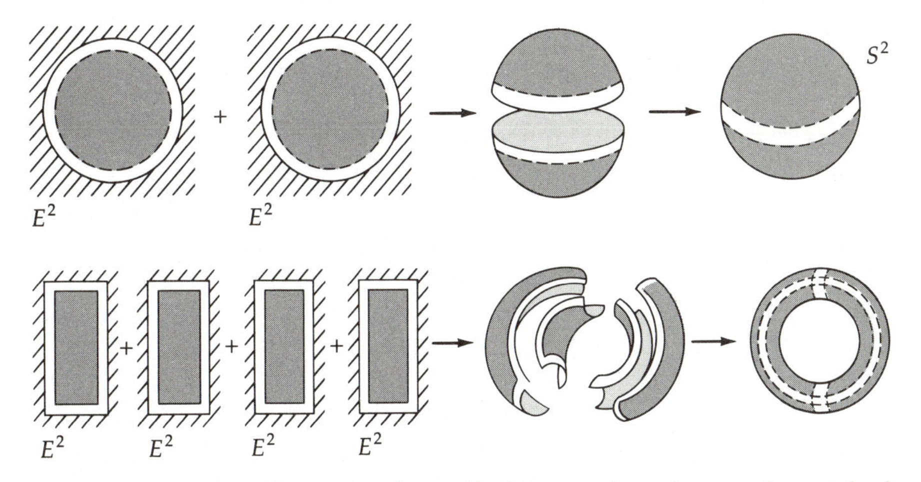

Figure 2.18 A two-dimensional manifold is a surface that can be patched together from pieces of a two-dimensional euclidean space (E^2). The diagrams show how this procedure can be applied to produce a sphere (*top*) and a torus (*bottom*), simply by specifying how the euclidean pieces should be glued together.

the space of positions of three balls on a billiard table is six-dimensional. If the balls are moving, then two more numbers (say, the speed and direction) are needed to specify the velocity of each ball. That adds up to a 12-dimensional space. Newton's laws of motion, in the form of differential equations, provide a way of describing how such a system of moving balls evolves within this 12-dimensional space.

Just as a wildflower guidebook highlights key features such as color and number of petals to help readers distinguish one plant from another, special manifold characteristics, often expressed as numbers or algebraic expressions, help mathematicians tell manifolds apart. Such expressions, known as topological invariants, provide a convenient way of putting manifolds into different categories.

Dimension, the number of coordinates required to specify a point in a given space, is an example of a topological invariant. It's the first level of classification in the world of mathematical manifolds. Manifolds may also be either bounded or unbounded. A circle is an example of a bounded, one-dimensional manifold, whereas a line stretching off indefinitely in both directions is clearly unbounded. The same distinction applies to spaces of any dimension.

Certain types of two-manifolds can also be classified in terms of an invariant called the *Euler number*. Imagine cutting a bounded two-manifold into triangles, being careful not to allow the vertex of one triangle to come from the middle of an edge of another triangle (see Figure 2.19). The number of vertices minus the number of edges (arcs) plus the number of regions (triangles) is defined as the surface's Euler number. Remarkably, that number doesn't depend on how you cut up the surface.

It turns out that each even integer less than or equal to 2 is the Euler number for exactly one two-manifold. The sphere has Euler number 2; the torus (analogous to an inner tube or the surface of a doughnut) has Euler number 0; the surface of a two-handled soup tureen has an Euler number of -2. In such a scheme, the surface of a sphere or a lump of clay falls into one group whereas the surface of a doughnut or a coffee mug falls into another. That's the

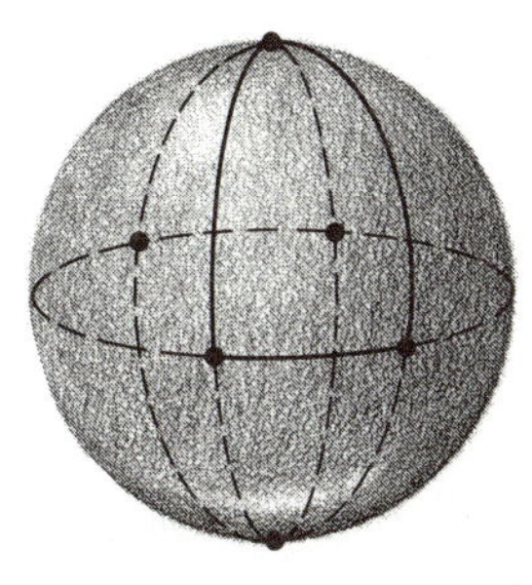

Number of vertices, $V = 6$
Number of arcs, $E = 12$
Number of regions, $F = 8$
$V - E + F = 2$

Figure 2.19 One invariant for identifying certain types of two-dimensional manifolds involves dividing a closed surface into several regions by marking a number of vertices on the surface and joining them by curved arcs. Using the formula $V - E + F$, where V is the number of vertices, E the number of arcs, and F the number of regions, produces a number that identifies the manifold. For a sphere and any other figure topologically equivalent to a sphere, such as a polyhedron, the answer is always 2. For a doughnut's surface, or torus, the answer is zero.

basis for the old joke that a topologist is someone who can't tell the difference between a doughnut and a coffee mug.

Botanists can place a particular plant first in a family, then into a genus, and into a species, making finer distinctions at each step. Similarly, topologists, using appropriate invariants, can examine in greater detail what manifolds look like and how one can be transformed into another. Much manifold study concerns the search for more finely tuned invariants that make increasingly subtle distinctions. Because manifolds in higher dimensions are impossible to visualize, these invariants often stand in for the manifolds themselves.

Mathematicians have developed reasonable, workable schemes for studying manifolds in every dimension except three and four. Dimension three remains a puzzle because proposed classification schemes cover only a portion of all conceivable three-manifolds. Recent attempts to demystify dimension four reveal it to be a special case with characteristics quite unlike those of any other dimension.

The central problem in classifying four-manifolds concerns the technical distinction between topological manifolds and smooth, or differentiable, manifolds. A ball, for instance, has a smooth, continuous surface. A closed, empty box has a continuous surface, but because it has sharp edges and corners, its surface isn't smooth.

The difference is crucial because topologists have ways of mathematically smoothing any sharp edges, creases, and jagged features in dimensions one, two, and three. For example, the surface of a box can be smoothed out in a reasonable way that transforms it into a sphere. In other words, there's no difference in these dimensions between topological manifolds (the more general category) and smooth manifolds. In five dimensions and higher, manifolds come in both the smooth and crinkly varieties, and mathematicians have a good understanding of when and how the different types occur. In four dimensions, the distinction between smooth and crinkly manifolds is much more complicated and difficult to sort out.

A similar distinction applies to transformations designed to test whether manifolds belong to the same class or fit into different categories. Like a lump of pizza dough, a topological manifold is free to be kneaded and distorted. In general, two such manifolds can be regarded as equivalent if one can be transformed into the other without tearing. Such transformations may involve a smooth transition or follow a crinkly course. Calling specifically for a smooth transition is a more stringent condition than simply showing that two manifolds are topologically equivalent, somewhat similar to a botanist determining to which subspecies a particular flower belongs.

The first major step in classifying four-manifolds was a proof that certain types could be identified on the basis of algebraic invariants called quadratic forms. Michael Freedman took 7 years to crack the problem. His 1981 proof showed that such manifolds can be constructed from simple building blocks and classified entirely on the basis of their quadratic forms.

Freedman's remarkable work unearthed many new examples of four-manifolds and established previously unknown transformations between known manifolds. But it didn't exclude the possibility that some of his four-manifolds may have creases that can't be removed in any way. In other words, dimension 4, unlike dimensions 1, 2, and 3, may contain manifolds that are topologically but not smoothly equivalent.

In 1982, Simon Donaldson proved that dimension four has just such an unexpected feature. He showed that not all topological manifolds can be constructed in a smooth way. Furthermore, for some manifolds, no amount of tugging or pushing rids the manifold of all its creases.

Donaldson used mathematical tools provided by theoretical physics to prove his point. He worked with a complicated set of mathematical expressions known as the Yang–Mills equations, which had proved crucial in physics for predicting the existence of new fundamental particles more massive than the electron. The Yang–Mills equations are notoriously difficult to solve. To find answers, physicists use the known geometric or topological properties of four-dimensional space to get information about potential solutions of the equations.

Donaldson took an even more difficult and daring route by starting with what little was known about solutions to the equations. He used that knowledge to extract information about the underlying four-dimensional space, essentially considering the solutions themselves as mathematical objects to be manipulated according to rules he derived. By showing how to compute quadratic forms from solutions of the Yang–Mills equations, Donaldson demonstrated that quadratic forms are not sufficient to distinguish between topological manifolds that are smooth and those that are not. Donaldson subsequently refined his methods to develop new, subtler invariants that distinguish between smooth manifolds even when they have the same quadratic form and are therefore topologically equivalent.

Donaldson also demonstrated that even when four-manifolds can be smoothed out, the process can take many different routes, leading to vastly different results. Other mathematicians extended Donaldson's work and took it in new directions. They discovered, for example, that four-dimensional, euclidean space can be given innumerable smooth descriptions. In other words, there exist exotic four-manifolds that are topologically but not smoothly equivalent to standard, four-dimensional euclidean space. In all other dimensions, euclidean spaces have a unique smooth description, which mathematicians have long used and understand well.

These startling discoveries pose equally perplexing riddles for mathematicians and physicists. In particular, four-dimensional space is where Einstein's theories must work and where modern physics resides. The existence of exotic four-dimensional spaces suggests there may be something different or special about four-dimensional space-time.

One striking feature of exotic four-dimensional spaces is their increasing complexity as their scale increases; they become infinitely complex at infinity. Such a picture may apply to the distribution of matter in the universe. Indeed, as astronomers study the universe on larger and larger scales, some see hints that the distribution of galaxies and interstellar matter doesn't seem to even out. They detect evidence for irregular arrangements of giant structures, punctuated by large gaps, as far as the aided eye can probe. Astronomer R. Brent Tully, in mapping the distribution of galactic clusters at cosmological distances, proposes that the local supercluster in which our Milky Way galaxy resides is actually part of what he calls the Pisces-Cetus complex, a gargantuan grouping of superclusters extending more than a billion light-years.

Tully's observations lie at the limit of observational work in astronomy, making measurements tough to interpret. Moreover, it is difficult to imagine how such an immense, irregularly distributed collection of galaxies could have formed through gravitational effects in the comparatively brief time available since the big bang (see Figure 2.20). Nevertheless, one might expect a universe like the one Tully sees, which becomes increasingly inhomogeneous as more distant regions are explored, if it were embedded in an exotic four-dimensional space rather than a space of the conventional euclidean variety. Like explorers who concentrate on their immediate surroundings and miss the distant mountains, physicists may find that their theories, formulated for only a small, well-behaved piece of space-time, may not work on a larger scale.

Figure 2.20 The existence of clusters of galaxies may be one clue that the universe has an exotic four-dimensional geometry. This photograph shows Stephan's Quintet, a cluster of five galaxies in the constellation Pegasus.

Mathematicians, too, are puzzled by what all this means. Four-dimensional space is indeed strange, with many mysteries yet awaiting solution. For instance, picturing exotic four-dimensional spaces, or four-spaces, remains a problem. Topologists now know exotic four-spaces are conceivable, but they don't know what they look like or how to build these extremely convoluted structures explicitly.

Furthermore, mathematicians are far from completing the task of classifying smooth four-manifolds. The new Donaldson invariants are difficult to define and compute. Topologists don't really understand how they work. No one knows how many are needed, and although the invariants work in specific cases, mathematicians have no idea how good they are at distinguishing manifolds in general.

Researchers also have two different pictures of four-manifolds—one in terms of a construction procedure and the other in terms of Donaldson's mysterious invariants. But they haven't yet succeeded in bridging the gap between the two. It's clear that any definitive classification scheme will have to be both subtle and intricate.

The methods used for studying four-manifolds show paradoxically that the study of space in five and higher dimensions is simpler and more amenable to understanding than is the study of space in three and four dimensions. Presently, efforts to advance the understanding of four-manifolds appear stalled. Perhaps further progress will follow more exchanges between physics and mathematics. Donaldson's work rested upon deep connections between mathematics and physics. Physicist Edward Witten's recent research linking quantum field theories and knot theory may be a step in the right direction.

3

Fitting Arrangements

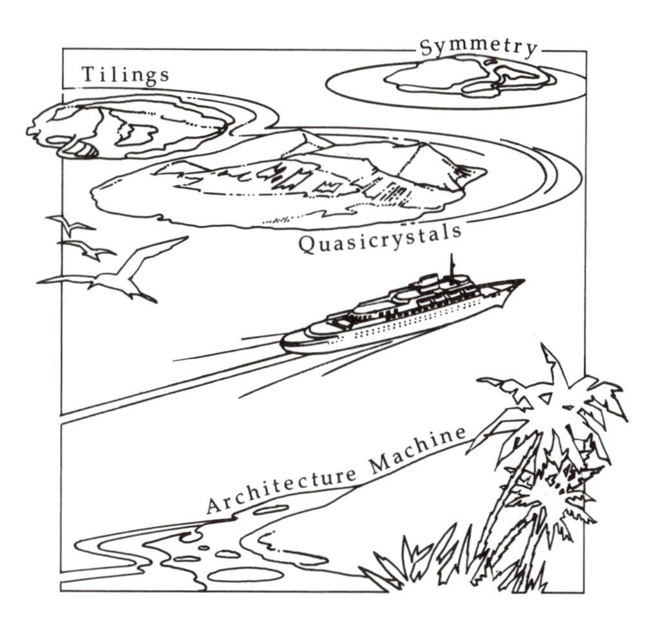

Decades ago, mathematician G. H. Hardy declared: "A mathematician, like a painter or a poet, is a maker of patterns. If his patterns are more permanent than theirs, it is because they are made with ideas." Mathematicians draw their patterns from many sources. Some are suggested by natural forms: the spiral distribution of seeds at the center of a sunflower or the delicate formations in a cluster of soap bubbles. Others arise out of practical human needs: arranging tiles to decorate a room or fitting bricks to build a wall.

Repeated patterns, with their captivating harmony and geometric logic, have long inspired both artists and mathematicians. From the simplest tiled floors and decorations on pottery to the enormously rich and intricate ornamentation on Islamic mosques, such patterns have great appeal (see Figure 3.1). The increasing popularity in recent decades of puzzles and games based on the interplay of shapes and positions further attests to the attraction of geometric forms and their relation to one another.

The study of tilings, in which geometric figures fit together perfectly to cover a flat surface, or a plane, is

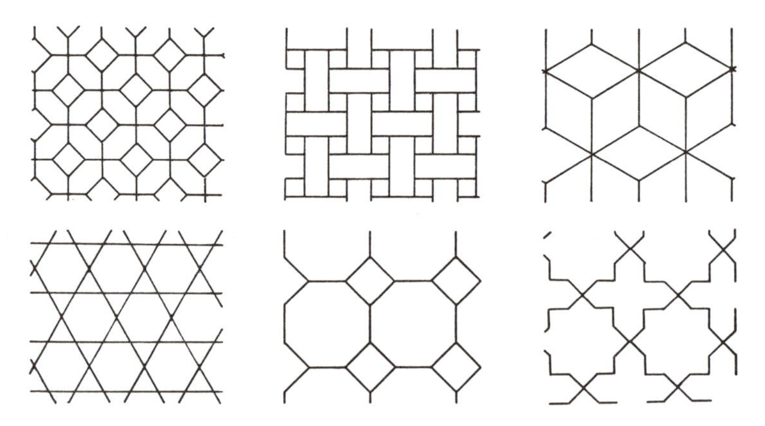

Figure 3.1 Different shapes of tiles can be used to generate an incredible variety of patterns. The tilings shown appear frequently in European and Middle Eastern architecture and designs.

becoming a powerful tool for exploring the frontiers of mathematics and science. And despite their seemingly childish ingredients, the problems of plane tilings are a genuinely difficult area of mathematics. Much remains to be learned.

"Perhaps our biggest surprise . . . was that so little about tilings and patterns is known," Branko Grünbaum and G. C. Shephard wrote in the introduction to their 1986 book *Tilings and Patterns*. Although tiling problems have roots that go back to antiquity, the authors discovered that "some of the most exciting developments in this area . . . are not more than twenty years old."

Paving the Plane

Part of what makes the study of tilings so appealing is that almost anyone can grasp some of the basic ideas of how they work. Snugly fitting together the pieces of a jigsaw puzzle or arranging tiles to fill a desired space is one of the earliest games of childhood. Such problems, whether in games and puzzles or in construction and remodeling projects, continue to hold the attention of adults.

Tiling is the process of fitting together flat geometric shapes so that pieces cover the plane without overlapping one another or leaving any gaps. It can also be expressed as the problem of dividing up the plane into shapes of a specified form. The result is a kind of giant jigsaw puzzle that stretches off to infinity.

The most basic question that anyone can ask is what tile shapes will completely fill an infinite plane. The most general answer to that question isn't known. Considering just the intricate plane-filling designs that artists have over the centuries worked into mosaics, carpets, fabrics, and baskets shows how complicated a catalog of all possible patterns would be.

Mathematicians and others interested in tilings have found partial answers by putting restrictions on the shapes of constituent tiles. For example, one can specify that the tiles be regular polygons, such as squares or equilateral triangles. In those closed figures, all sides are equal in length and all internal angles are the same size. It's easy to see that any square or equilateral triangle can tile the plane. The result is a grid of squares or a network of equilateral triangles. Regular hexagons also work, as seen in chicken wire and bathroom tiling. Honeybees exploit this arrangement when they build strong, efficient storage units in the form of honeycomb for their hives. A hexagonal tiling pattern displays no continuous straight lines, meaning that a floor made up of tiles in such a pattern would presumably be stronger and more resistant to cracking (see Figure 3.2). Squares, equilateral triangles, and hexagons are the only regular polygons that tile the plane.

The next step is to consider patterns in which two or more regular polygons are fitted together corner to corner in such a way that the same polygons surround each vertex. For example, an otherwise nontiling octagon will fit the scheme if small squares are used to fill the gaps. In fact, precisely eight such patterns exist, made up of different combinations of triangles, squares, hexagons, octagons, and dodecagons (see Figure 3.3). All would make pleasing flooring patterns. A simple change in the rules, such as

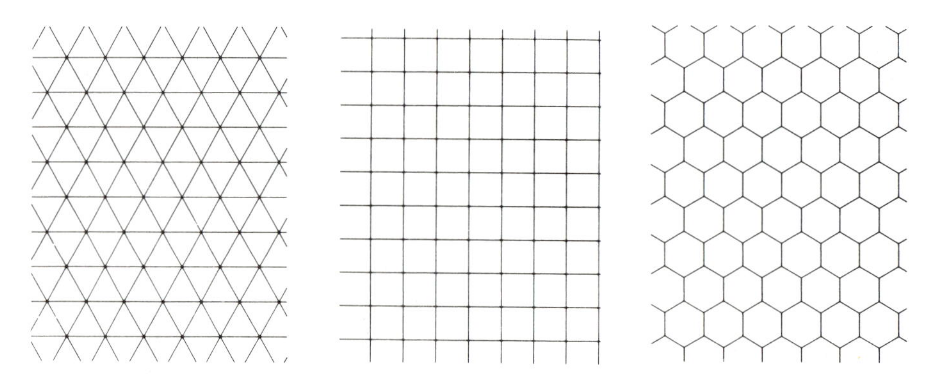

Figure 3.2 Only three types of regular polygons—equilateral triangles, squares, and hexagons—can be fitted together corner to corner to cover the plane. Unlike the square and triangular grids, the hexagonal grid has no continuous straight lines that cut across the entire pattern.

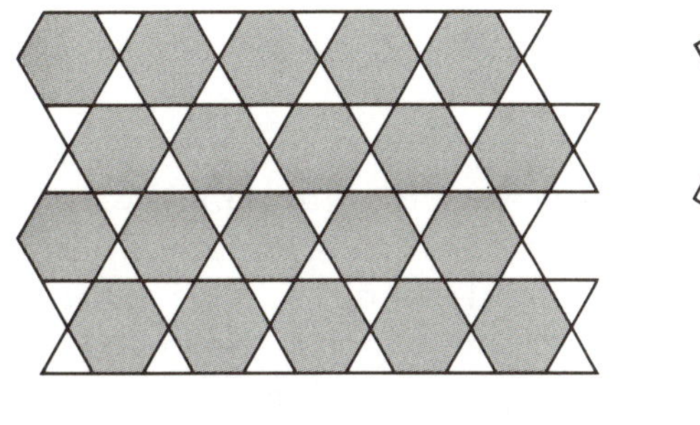
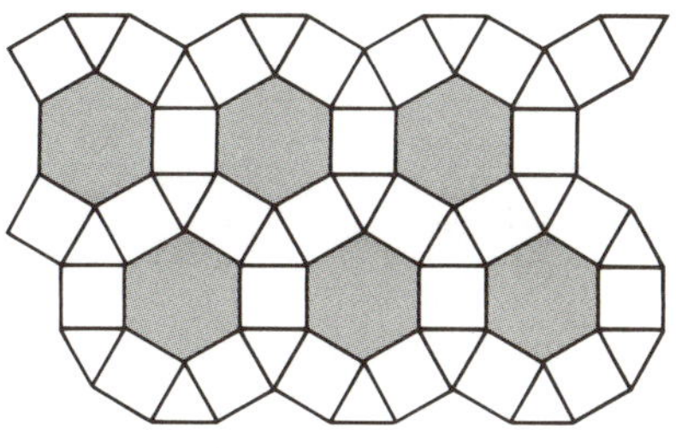
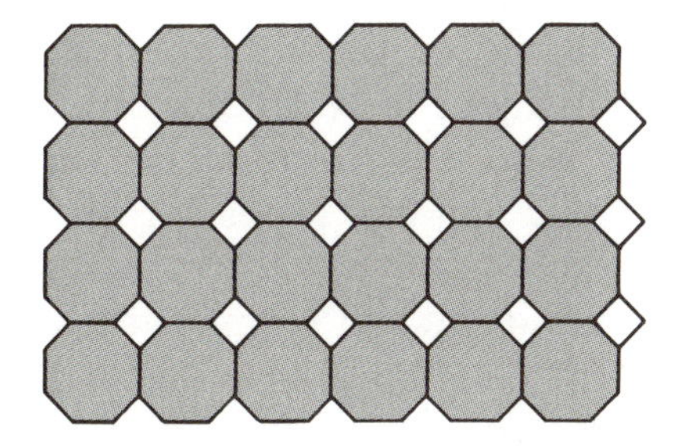

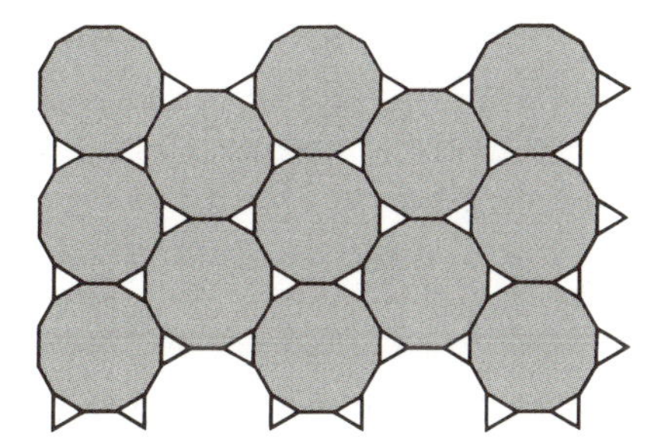
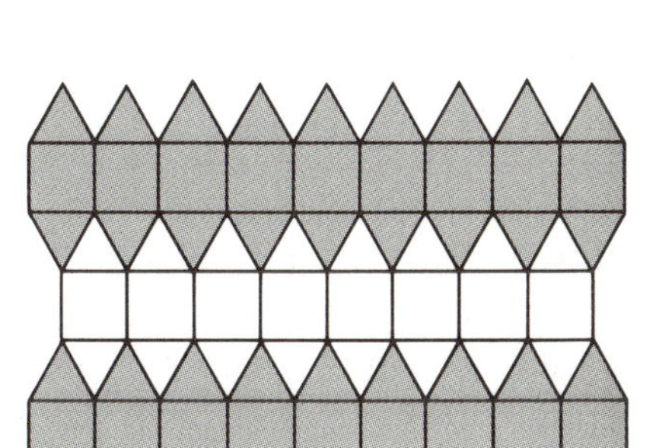
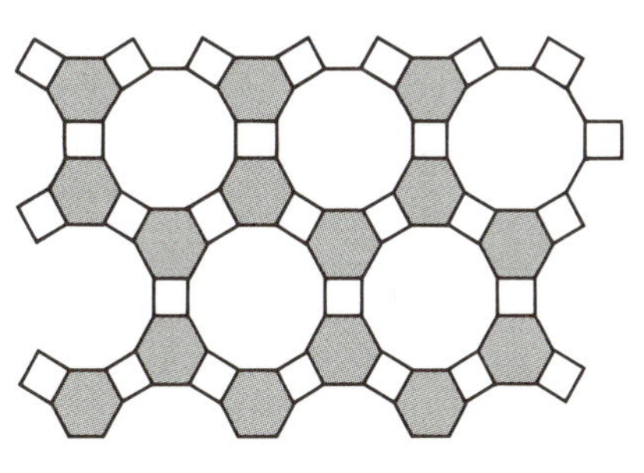
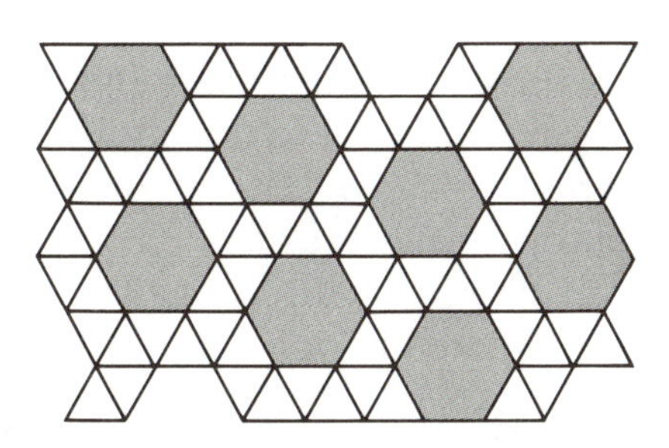

Figure 3.3 In semiregular tilings, two or more kinds of regular polygons are fitted together corner to corner in such a way that the same polygons, in the same order, surround every vertex. There are exactly eight of these tilings, made up of different combinations of triangles, squares, hexagons, octagons, and dodecagons.

removing the restriction about arrangements surrounding a vertex, allows the creation of an infinite variety of mosaics.

The possibilities are endless. What if tiles are made up of a certain number of squares, equilateral triangles, or hexagons stuck together to form a single, larger unit, which is then used as a tile? What if pairs of different figures are used? What if the polygons don't have to be regular? The list goes on and on.

One way to help keep track of some of the innumerable possibilities for tiling the plane is to consider the symmetry of the resulting pattern. Symmetry has long held a fascination for the human mind. Ancient pottery exhibits regularly repeating patterns. People admire the symmetry found in nature, from the arrangement of flower petals to the regular crystal forms evident in diamonds and other minerals.

The key elements of symmetry are regularity and balance. There's symmetry in a checkerboard design but not in random paint splashes. Rotating a five-armed starfish one-fifth of a turn doesn't change its appearance, although the positions of individual arms have changed. The starfish has a fivefold symmetry, and a flower with three petals has a threefold symmetry. The most symmetrical object imaginable is the sphere, and nature abounds with spheres: stars, planets, water droplets, soap bubbles. A perfect sphere looks the same from any angle.

Symmetry also enters into physical phenomena. A methane molecule has one carbon at its center and four hydrogen atoms arranged at the corners of a tetrahedron surrounding the central atom. The molecule's symmetries determine how it vibrates (see Figure 3.4). In more complicated molecules, measurements of vibrations provide information about the molecule's structure and its symmetries.

Basic maneuvers called symmetry operations provide a basis for classifying objects or patterns in terms of symmetry. A grid of squares shows no change when reflected in a mirror, or rotated 90 degrees, 180 degrees, or 270 degrees. Shifting the pattern sideways also leaves the pattern's appearance unchanged. Thus, a square pattern has a high degree of symmetry. Many tilings are periodic, meaning that a basic unit consisting of one or more tiles is regularly repeated throughout the entire pattern.

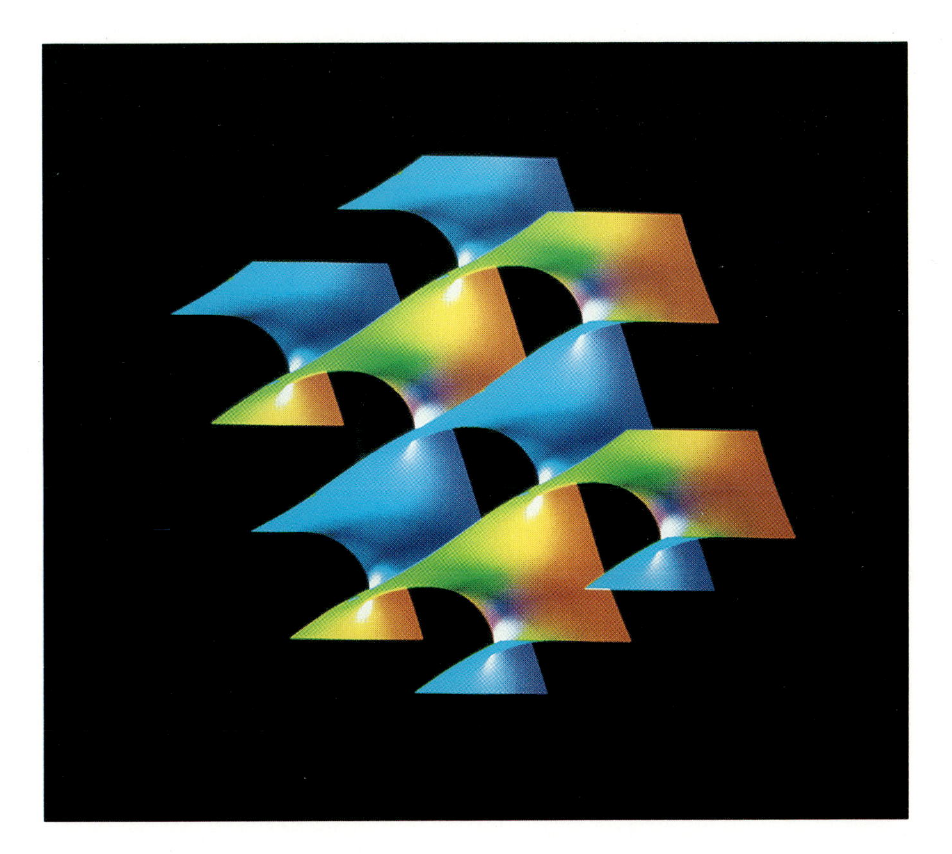

1 Saddle Grid

2 Etruscan Venus

3 Ida Thoughts

4 A Ring of Horses

5 The Heart of a Sphere

6 Knot Math

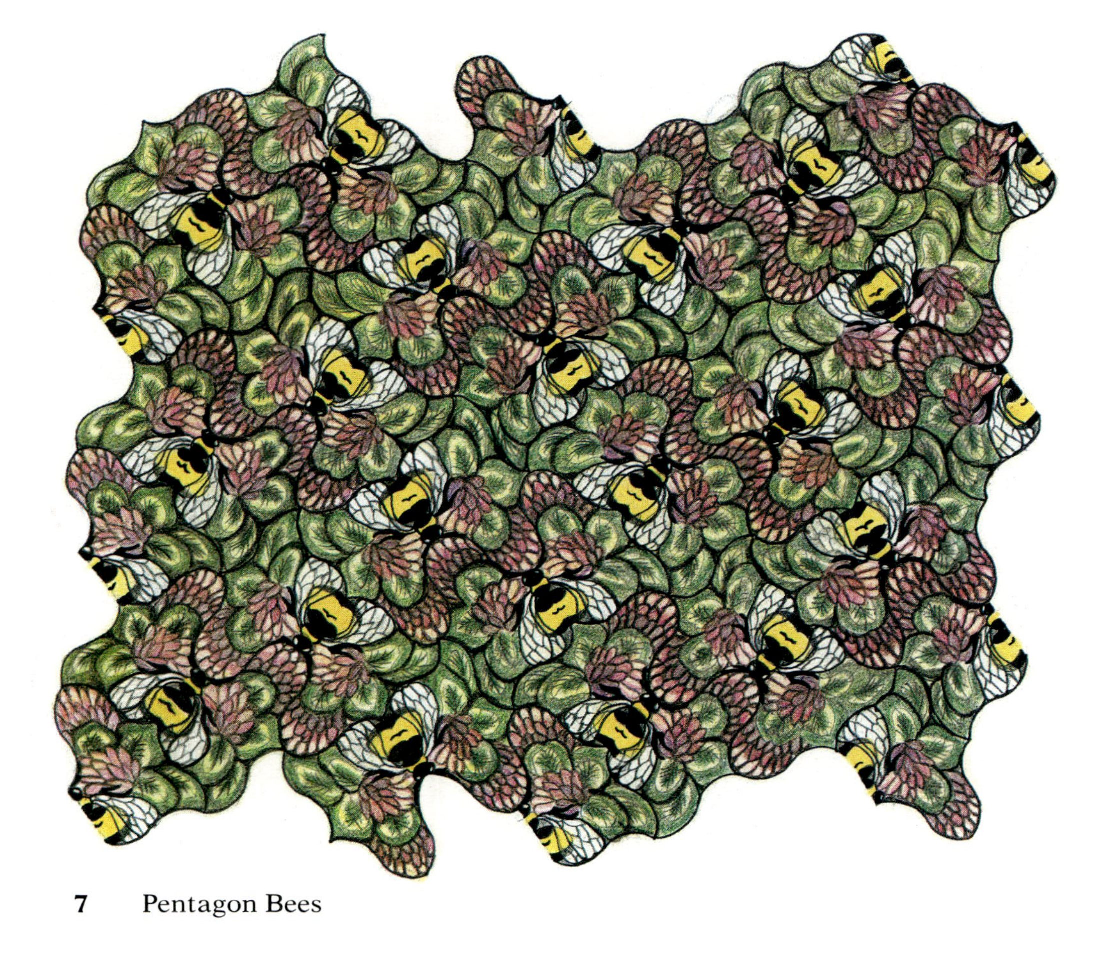

7 Pentagon Bees

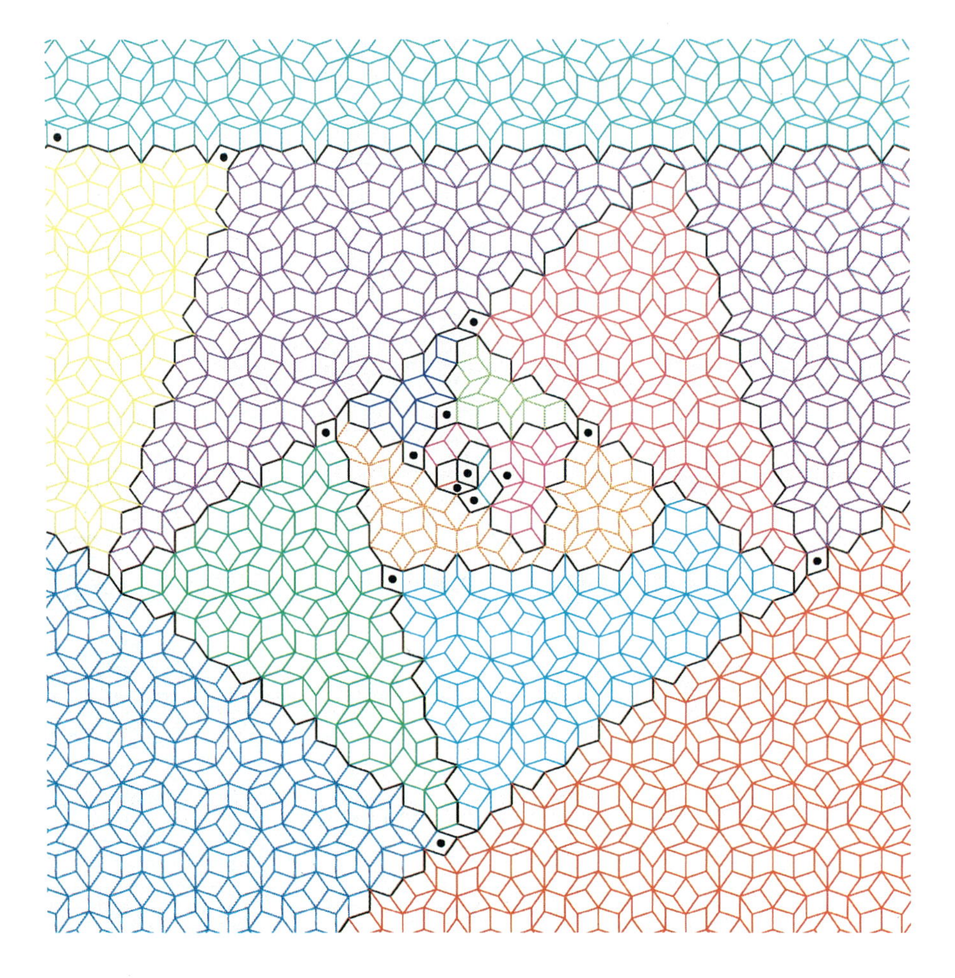

8 Growing to Infinity

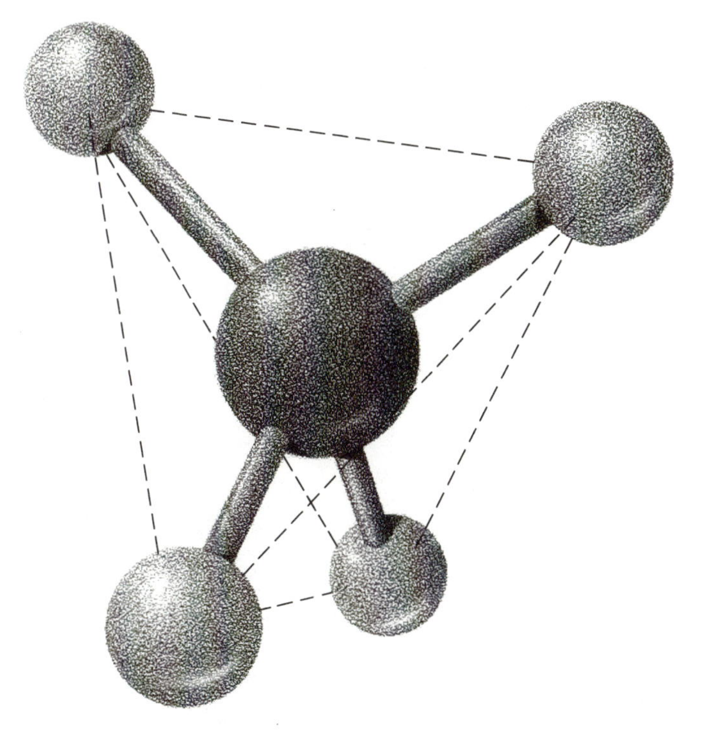

Figure 3.4 A methane molecule has a tetrahedral symmetry, with a hydrogen atom at each of a tetrahedron's four corners and a carbon atom at its center. Such a molecule vibrates in characteristic ways.

It turns out that all tilings that cover the plane with a repeated pattern belong to a set of 17 different symmetry groups that exhaust all of the fundamentally different ways in which patterns can be repeated endlessly in two dimensions (see Figure 3.5). Artisans in a variety of cultures have exploited practically all of these possible repeated patterns, from the intricate decorations on Islamic buildings in thirteenth-century Spain to the fascinating patterns painted on African pottery.

The same idea can be extended to three-dimensional objects. Crystallographers in the nineteenth century established the exis-

Figure 3.5 Examples of tilings for each of the seventeen crystallographic, or wallpaper, symmetry groups.

tence of 230 repeating patterns in three dimensions. Their catalog became a valuable tool for classifying minerals and other crystals and studying their internal structure (see Figure 3.6).

Now scientists and mathematicians have carried the notion of symmetry beyond geometric patterns. Physicists talk of symmetry in the physical laws governing the behavior of matter, for example. The concept is an extension of the idea that anything, whether a geometric object, an algebraic expression, or a physical law, is symmetrical if there is something we can do to it so that after we have done it, it looks the same as it did before. For instance, the algebraic expression x^2 is symmetrical in a particular way because it doesn't matter whether the number substituted for x is positive or negative. The answer will be the same regardless of the number's sign.

Physical laws can have a variety of possible symmetries. Physicists can ask whether the law is the same everywhere in space, whether it stays the same over time, whether it works in a mirror-image universe, whether it applies if time is reversed, whether it gives the same results when electrical charges are exchanged (matter exchanged for antimatter), and so on. Physicists have

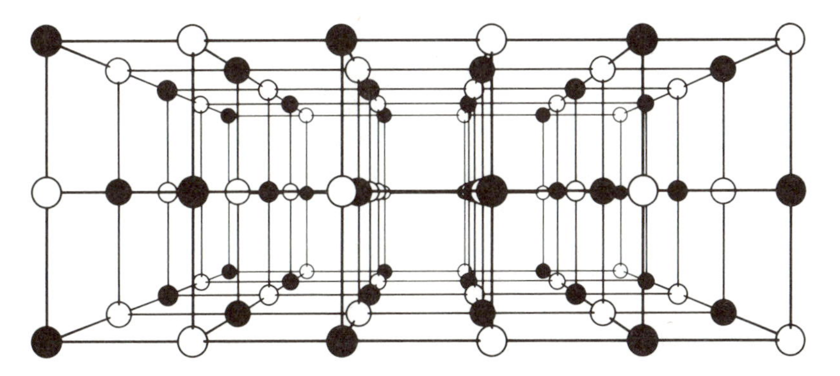

Figure 3.6 Crystals of common table salt (sodium chloride) form perfect cubes. Such a symmetrical shape is a manifestation of the material's internal structure, reflecting how individual sodium ions (*filled circles*) and chloride ions (*open circles*) are packed together in a regularly repeated, cubic lattice.

made considerable progress in their understanding of matter and its interactions by considering symmetries in physical laws, striving to find the most symmetrical possible way of expressing nature's laws, yet always looking out for "broken" symmetries, which signal exceptions to the rule.

The mathematical language of symmetry is *group theory*. Group theory describes in abstract terms the relations between different symmetries of the same object. It applies to descriptions of the shape of a starfish or a crystal, to mathematical equations and geometric forms, to physical phenomena such as the vibrations of a methane molecule, and to fundamental physical laws. Concepts from group theory pervade modern mathematics and supply physicists and chemists with an essential tool for understanding their theories and experiments.

In general, a group consists of a particular set of symmetry operations. For example, an equilateral triangle can undergo 6 different motions that leave it looking exactly the same as before the operation (see Figure 3.7). Each operation is a symmetry of the triangle. The set of all 6 symmetries is called the equilateral triangle's symmetry group. Because the group contains 6 symmetry operations, it is said to be of order 6. Similarly, a square has 8 symmetries, and a regular pentagon has 10 symmetries.

Groups come in many different guises. Any mathematical object, whether an equation or a geometric shape, can possess symmetries; that is, can undergo certain transformations that preserve the object's basic structure. In the case of an equation, the mathematical object is the set of all solutions of the equation. Group theory expresses all the different ways in which these solutions can be transformed and rearranged among themselves. The structure that's preserved is the system of algebraic relations among the solutions.

Group theory was invented at the beginning of the nineteenth century by Évariste Galois, a young French mathematician who wanted to prove that certain equations couldn't be solved by simply using a formula to get the answer. His scheme worked because whether an equation can be solved depends on the equation's symmetries.

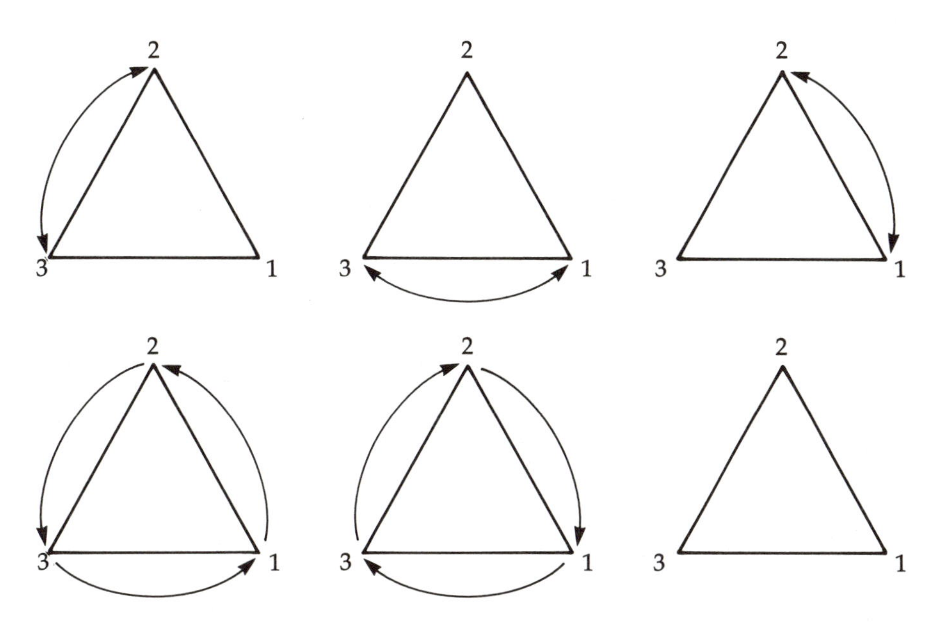

Figure 3.7 A symmetry of an equilateral triangle is a way of moving the triangle so as not to disturb its appearance. For example, leaving vertex 1 alone and flipping the triangle over so that vertex 2 is interchanged with vertex 3 is one of the symmetry operations possible for an equilateral triangle (*top left*). There are six such operations, including one in which the triangle is left alone (*bottom right*).

Mathematicians have grown accustomed to thinking about the symmetry of overall patterns, but many interesting problems can't be solved in terms of symmetries and group theory. Instead, it's useful to look at the relation of one part to another, usually to its nearest neighbors. One such problem concerns the possibility of devising suitably shaped tiles that cover the plane yet don't repeat in a regular pattern. Such a design may have a high degree of order but no symmetry group. A Penrose tiling, discussed in the next section, is one example.

Similar problems of deciding whether tiles of a given shape lead to a tiling of the plane arise with tiles in the form of convex polygons. A polygon is convex if no two points within it can be connected by a line that has to go outside the polygon. A regular hexagon is a convex tile; a five-pointed star is not (see Figure 3.8). Regular hexagons, like those in a honeycomb, cover the plane, but many types of "irregular" hexagons do not. In 1918, Karl Reinhardt, a graduate student at the University of Frankfurt in Germany, settled the hexagon question once and for all by proving that only three types of hexagons work (see Figure 3.9). On the other hand, any triangle or quadrilateral, no matter how irregular, will tile the plane (see Figure 3.10). But no convex polygon having seven or more sides accomplishes the task.

Pentagons are a particularly interesting case. Tiles in the shape of regular pentagons leave gaps that can't be filled in. But any pentagon having a pair of parallel sides does work (see Figure 3.11). Reinhardt defined five classes of convex pentagons that tile the plane. Most people assumed his list was complete and thought the problem was closed—until 1968, when Richard Kershner found three more classes of plane-tiling pentagons that had somehow been overlooked. Martin Gardner discussed the problem in a 1975 Mathematical Games column, "On Tessellating the Plane with Convex Polygon Tiles," which resulted in the addition of

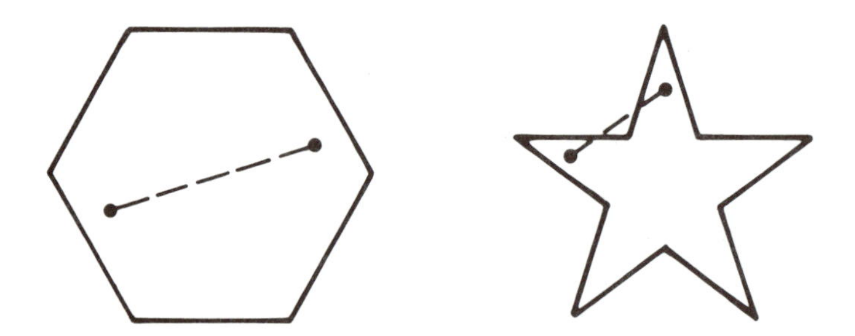

Figure 3.8 A hexagon (*left*) is an example of a convex polygon because any two points within it can be joined by a straight line that doesn't have to go outside the hexagon. In contrast, a five-pointed star (*right*) is a nonconvex polygon because there are pairs of points that must be joined by straight lines crossing the star's boundaries.

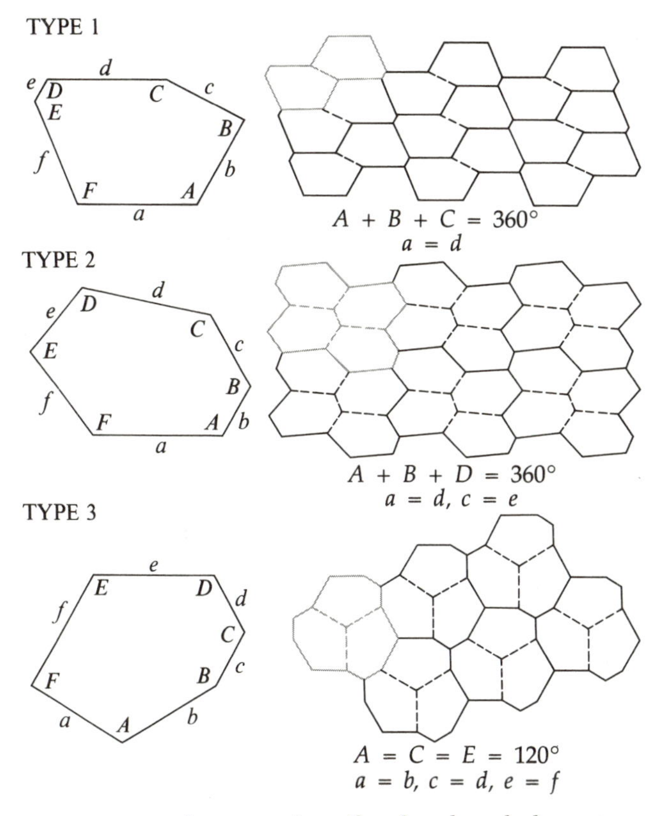

Figure 3.9 Any convex hexagon that tiles the plane belongs to one of three classes. The illustration gives an example of each type of convex hexagon and a portion of its tiling pattern. The gray lines outline a "fundamental region" that tiles the plane by translation.

another member to the family. Inspired by the same article and its follow-up, amateur mathematician Marjorie Fox found several more types of pentagons that fit snugly together (see Figure 3.12). She even developed her own system of notation to keep track of all the possibilities she investigated and turned some of her unique tilings into graceful pictures (see Color Plate 7).

Marjorie Fox's efforts raised the total to 13 distinct classes of convex pentagons that can be used to tile the plane. A fourteenth

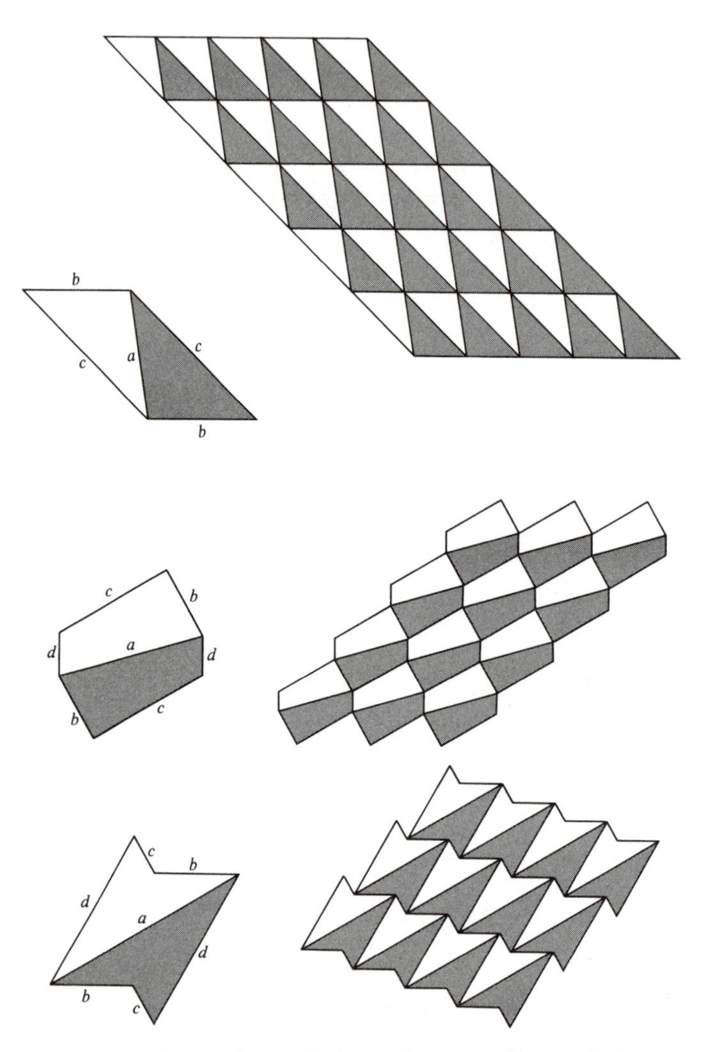

Figure 3.10 Triangles and quadrilaterals, regardless of shape, tile the plane.

was discovered in 1985. Is the list now complete? No one knows. A proof that the 14 known classes of pentagons cover every possible case has not yet been found.

Artists and designers don't normally use group theory to decide what patterns to make and how to construct them. They use

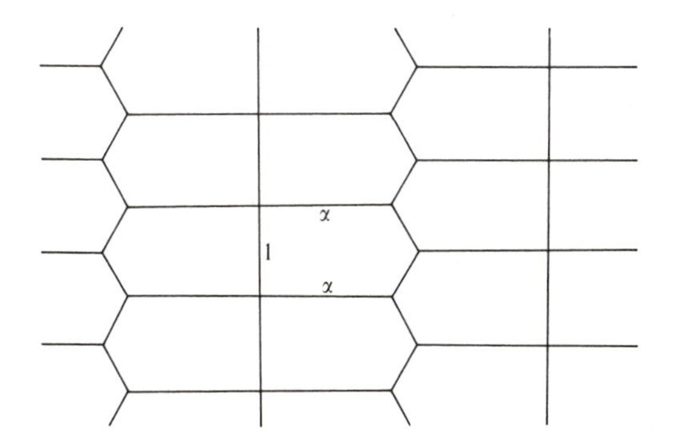

Figure 3.11 Convex pentagons with pairs of parallel sides tile the plane.

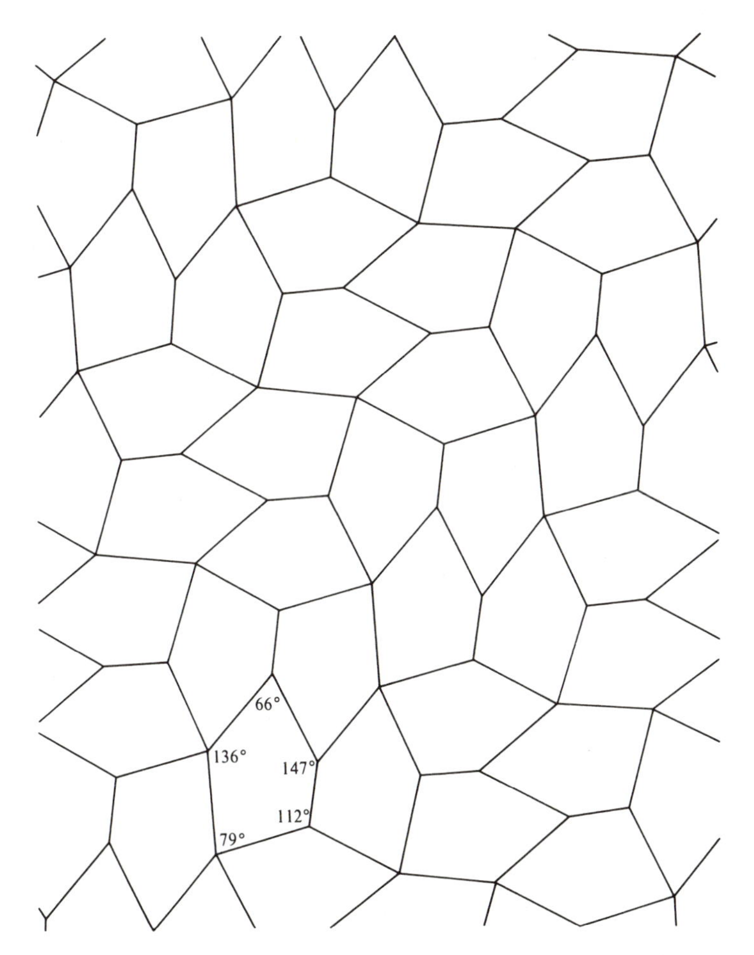

Figure 3.12 This tiling, which uses convex pentagons, was discovered by amateur mathematician Marjorie Fox and represents one of the 14 known classes of convex pentagons that tile the plane.

rules that depend far more on the relationships between neighboring tiles than on the overall symmetries of the resulting patterns. That approach suits the common practice of laying down one piece after another to build up the desired pattern. On the other hand, although artisans have a good idea of which patterns work, they are not always sure that certain patterns won't work. Intuitive methods lack the certainty that mathematics can provide. For a given set of rules, mathematicians can often work out all the possibilities and prove that no other examples could possibly fit that particular set of rules.

From Here to Infinity

In the 1970s, mathematical physicist Roger Penrose found he could assemble tiles shaped like fat and skinny diamonds into patterns that fill the plane yet don't repeat themselves at regular intervals. Like pentagons, the resulting tiling patterns have a fivefold symmetry. Unlike pentagons, the diamond tiles, when properly placed, leave no gaps (see Figure 3.13).

Initially just a playful creation, Penrose's tilings took on added significance with the unexpected discovery of crystalline materials that showed fivefold symmetry. Such a symmetry is physically impossible for a crystalline material in which groups of atoms fall into a regularly repeating, or periodic, pattern. The rules of geometry for a regular latticework in three-dimensional space allows only twofold, threefold, fourfold, and sixfold symmetries.

Penrose's tilings, with their intriguing blend of order and disorder, provide simple geometric models for how groups of atoms may be arranged within these novel materials, now known as quasiperiodic crystals, or more simply, quasicrystals. A quasicrystal doesn't have the traditional lattice structure of an ordinary crystal. A view from inside a quasicrystal shows that its lattice-

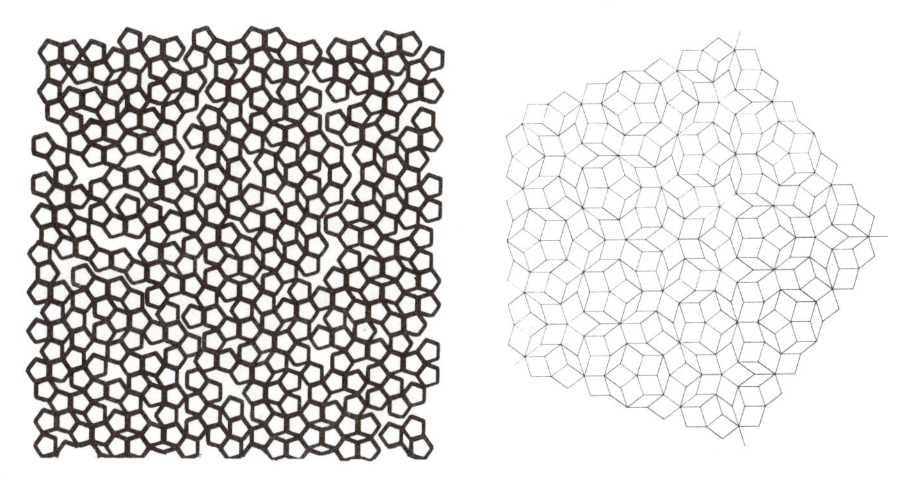

Figure 3.13 Whereas regular pentagons don't fit together to cover the plane (*left*), a set of fat and skinny diamonds produces a tiling pattern with a fivefold symmetry (*right*).

work looks somewhat different from place to place, even though its overall structure has a kind of long-range order.

However, the simple Penrose model of a quasicrystal has a serious flaw. Penrose's tiling scheme requires the diamond tiles to be fitted together in specific ways. Explicit edge-matching rules stipulate which sides and vertices are allowed to meet (see Figure 3.14). Otherwise, it would be possible to place the tiles so they fall into a regularly repeating pattern. But the rules aren't sufficient to guarantee that any number of tiles would fit together to create a flawless pattern. It's easy to place a tile properly yet run into trouble many moves later, ending up with a gap that tiles of neither shape can fill. That makes it hard to picture how atoms or molecules, influenced largely by their nearest neighbors, would have enough information to arrange themselves into flawless patterns on a large scale.

When tiling a floor with squares, you need look only at adjacent squares to know where the next tile goes. Ordinary crystals, made up of building blocks consisting of groups of atoms, grow by

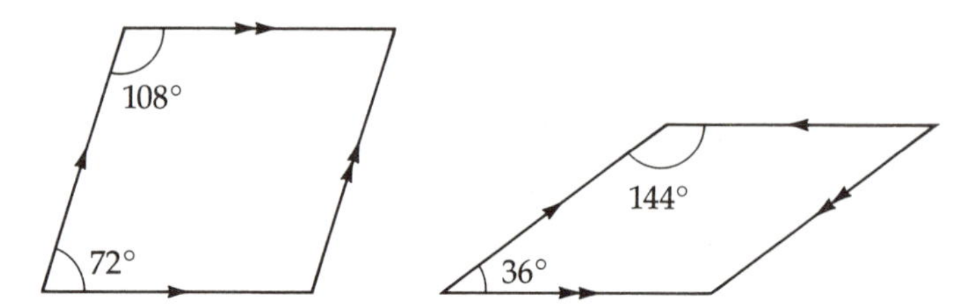

Figure 3.14 Fitting together the two types of diamond tiles pictured so that the arrows superposed on their edges match produces a Penrose pattern.

similar local rules. During crystallization, atoms have no way of checking what is going on elsewhere before positioning themselves. They fall into a regular pattern because the physical forces acting on corresponding atoms in different blocks are the same everywhere. If enough atoms end up in the wrong places, the crystal is defective, or fails to grow.

In a quasicrystal, each fivefold block sits in a slightly different environment, with an apparently different pattern of forces. Under such circumstances, how do different atoms conspire to aggregate in the right proportions and in the correct locations to maintain a degree of order over long distances? The lack of a set of local rules for generating perfect Penrose tilings led most quasicrystal researchers to consider alternative models for quasicrystalline materials. Many who had studied the Penrose tilings concluded that no local rules were likely to be found.

Indeed, Penrose himself proved that given only the matching rules, there was no step-by-step algorithm for placing tiles so as to create a perfect Penrose tiling. The rules guarantee a nonperiodic pattern, but not one free of defects. In other words, a worker, given a stack of Penrose tiles to cover a bathroom floor and no global plan on how to proceed, is bound to fail, eventually producing a flawed pattern with one or more spaces that can't be filled.

But rules don't have to be completely local to be relevant to a physical system. In 1987, George Onoda, a ceramics expert at IBM took up the challenge. By playing with a pile of about two hundred diamond-shaped tiles, Onoda found that, by eye, he could put together flawless structures using up all his tiles. He developed a list of empirical rules for generating perfect tilings. For example, fat tiles always seemed to end up in chains or rings. Skinny tiles were by themselves or in pairs.

Onoda demonstrated his scheme to quasicrystal theorist Paul Steinhardt and IBM colleague David DiVincenzo, who had experience in using computers to study tilings. Steinhardt and DiVincenzo helped Onoda convert his laundry list of visual rules into a simple statement about vertices. The result was a catalog of the eight different ways in which tiles can be allowed to meet at a vertex (see Figure 3.15). A new tile can be added to an edge of a given cluster of tiles only if each shared vertex matches one of the eight allowed arrangements. Tiles are placed first wherever there is only one possible way to complete a vertex not yet entirely surrounded. If more than one such placement is available, the "forced" moves can be done in any order.

Computer simulations of this procedure show that patterns evolve in a characteristic way. By adding tiles only where they are forced to go by the rules, a given cluster develops a faceted structure (see Figure 3.16). While growth proceeds along rough faces,

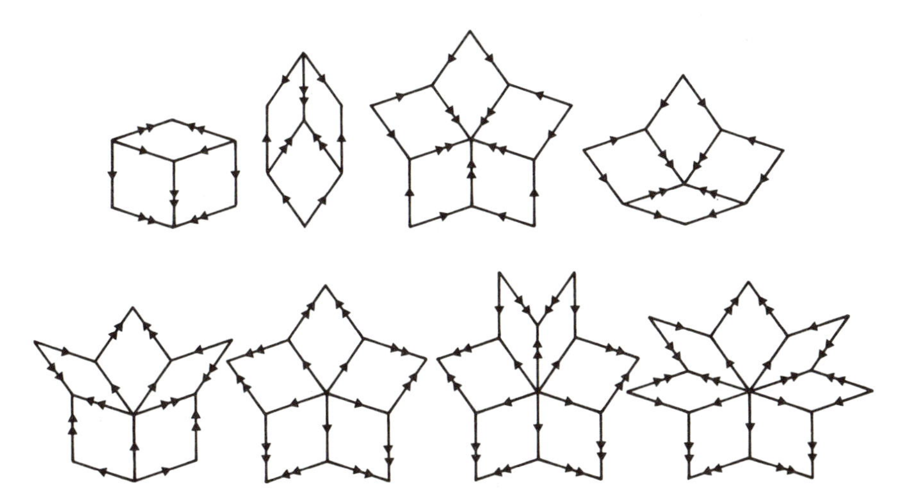

Figure 3.15 The eight possible ways in which vertices of tiling diamonds can meet to produce a perfect Penrose tiling.

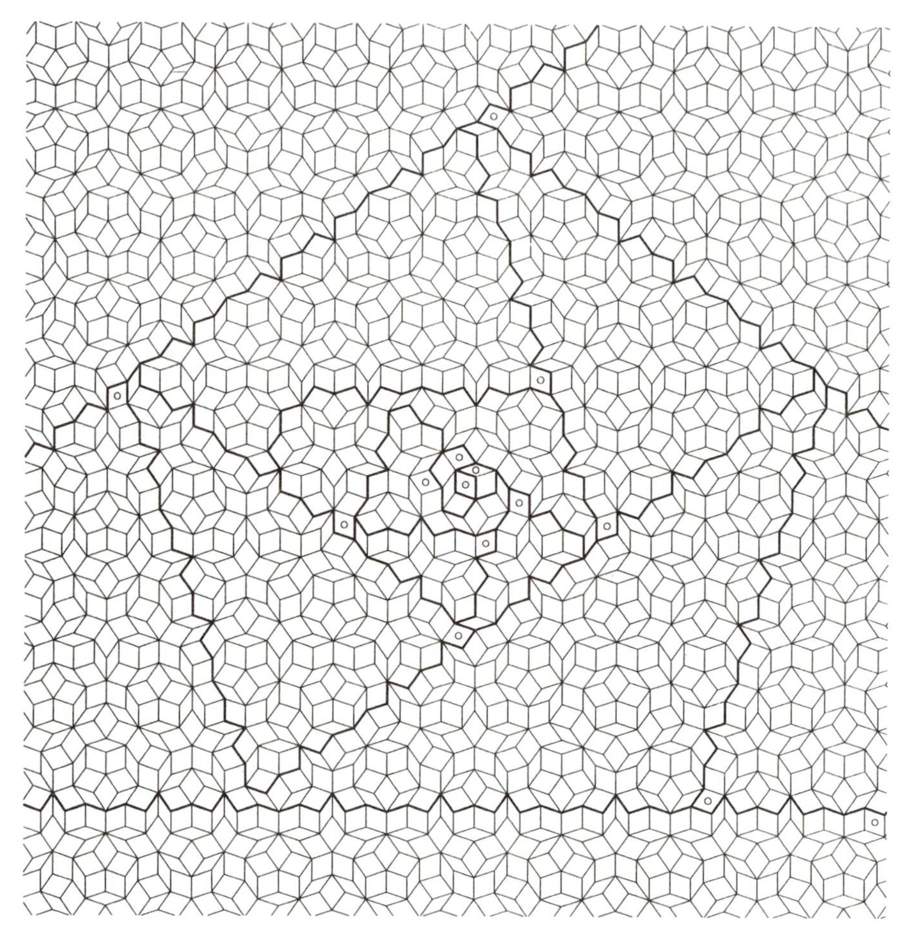

Figure 3.16 A succession of clusters grown from a seed consisting of a single tile. Tiles are added to the seed according to the vertex rules until there are no more forced moves. This produces a faceted "dead" shape. After placing a tile at a corner of the dead shape, growth proceeds until the next dead shape appears.

it stalls on a smooth face. Eventually, the whole pattern has smooth faces along its perimeter, and no more forced moves are available. The resulting pattern is known as a "dead" shape or surface.

Originally, the researchers proposed that if no forced moves are available, then a fat or a skinny tile, consistent with the vertex rule, can be added anywhere along a randomly chosen edge of the dead shape. However, because that method of restarting growth could sometimes lead to a defect, a new rule was required. The answer was to add a tile to one of the dead shape's corners. Then growth could proceed again. However, this change turns a strictly local procedure into one that requires examining the dead surface's entire perimeter to discover the right place to continue defect-free growth.

The modified process does generate arbitrarily large perfect tilings, but it proceeds in fits and starts. Growth is rapid until a dead surface forms. It stalls until the right starting point for a new growth spurt is found, then it continues very rapidly again until the next dead shape appears, and so on. Catalogs of the dead surfaces that occur during growth show that the same kinds of shapes appear over and over again on larger and larger scales (see Color Plate 8).

The results provide new insights into how materials with only short-range atomic interactions can grow into large, nearly perfect quasicrystalline grains. In ordinary crystal growth, some crystal surfaces are known to have sites that are "stickier" than others, encouraging atoms or molecules to settle there, rather than in other locations. If any quasicrystalline systems mimic the kind of growth rules now established for Penrose tilings, then these materials may have sticky and nonsticky sites corresponding to the forced and unforced moves in the Penrose model.

If that model holds, then the most common approach for growing quasicrystals—solidifying the molten alloy quickly, then slowly heating up the solid to try to get larger crystals with fewer defects—is inappropriate. Perhaps researchers should be quenching the molten material as slowly as possible to give atoms more time to find the "sticky" spots.

One problem with this model is that the stickiness of corner's would have to be significantly less than the stickiness of forced sites in a growing tiling pattern. But that means a long delay before a pattern starts to grow again after reaching a dead shape, when only corners are available for further growth. Real crystals form much more quickly than such a model would predict.

Another problem is that real crystal growth is unlikely to be so orderly. Defects are bound to occur. But this doesn't necessarily result in a totally disordered pattern. Computer simulations of the growth of Penrose tilings show that defects don't necessarily destroy the overall pattern. In some cases, the pattern repairs itself, growing around the defect and carrying on almost as if nothing had happened (see Figure 3.17).

Interestingly, researchers have discovered it's possible to build a perfect Penrose tiling if the pattern starts around a particular kind of defect: an unfillable void at the pattern's core. With such a defect, all moves are forced, and the tiling pattern proceeds flawlessly to infinity (see Figure 3.18). What that means physically is that if one such defect is present, the growing crystal will always have sticky points on its surface. If such a defect could be isolated, it would be an ideal feature around which to grow quasicrystals quickly. Computer simulations show that, in general, defect-based tilings grow much faster than perfect tilings.

A number of mathematicians and scientists also have taken up the task of generalizing the extended Penrose rules to other tilings with fivefold symmetry, to tilings with eightfold and twelvefold symmetries, and into three dimensions. So far, that task has turned out to be surprisingly difficult, although progress has been made.

In three dimensions, the pentagon's role is played by the icosahedron, a fivefold-symmetric solid with 20 triangular faces arranged so that 5 faces meet at each vertex. It's possible to pack a crate with cubes and leave no gaps, but because icosahedra don't fit snugly together, it can't be done with icosahedra. If groups of individual atoms happen to be arranged according to an icosahedral geometry, it would be impossible to construct a periodic latticework of such units.

But just as fat and skinny diamonds can be used to create a nonperiodic Penrose tiling with a fivefold symmetry, two rhombohedra, one thick and one thin, with identical facets, can be used to build up a three-dimensional, nonperiodic structure. These rhombohedra are the three-dimensional equivalents of Penrose's

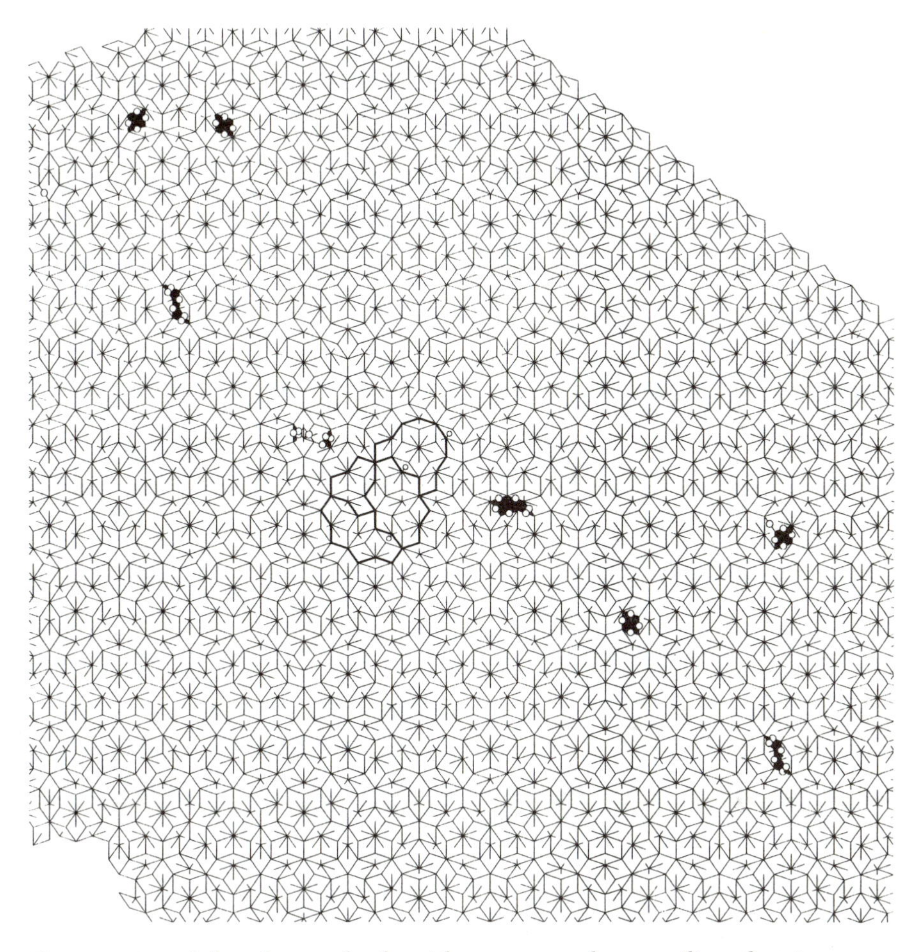

Figure 3.17 A local growth algorithm can produce a tiling that incorporates defects (*dark areas*) yet retains the semblance of a fivefold symmetry.

two types of diamond tiles (see Figure 3.19). The trick is to find rules ensuring that the rhombohedral tiles don't fall into a regularly repeating pattern. With appropriate rules, the resulting nonperiodic patterns would be icosahedrally symmetric in the same loose sense that the Penrose tilings are fivefold symmetric.

However, in sharp contrast to the simple matching rules used for placing diamond tiles, the rules for fitting together the rhom-

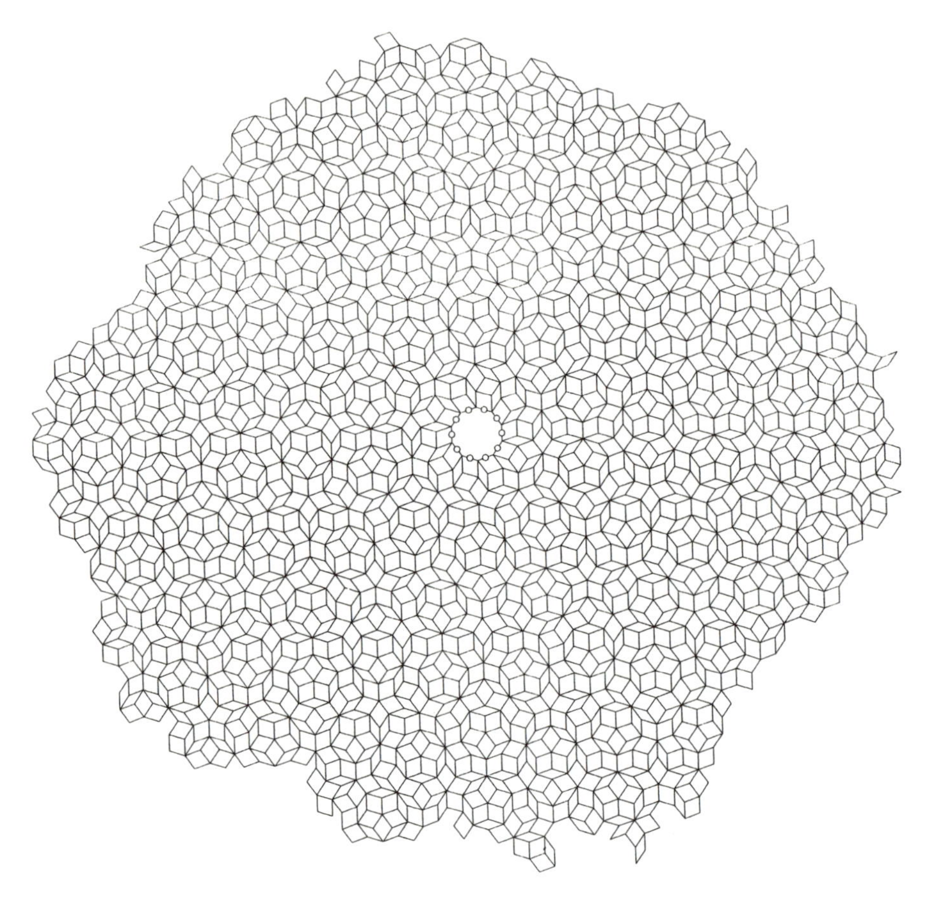

Figure 3.18 Certain pointlike defects known as decapods are ideal seeds for the growth of perfect Penrose tilings. If a cluster contains one of these defects, a dead shape never appears, and the tiles are forced throughout the plane.

bodedra are extremely complicated and difficult to define. Instead of just two sets of arrows, or decorations, researchers have found that to make the three-dimensional system work, the thick rhombohedron carries 14 markings, while the thin rhombohedron has 8.

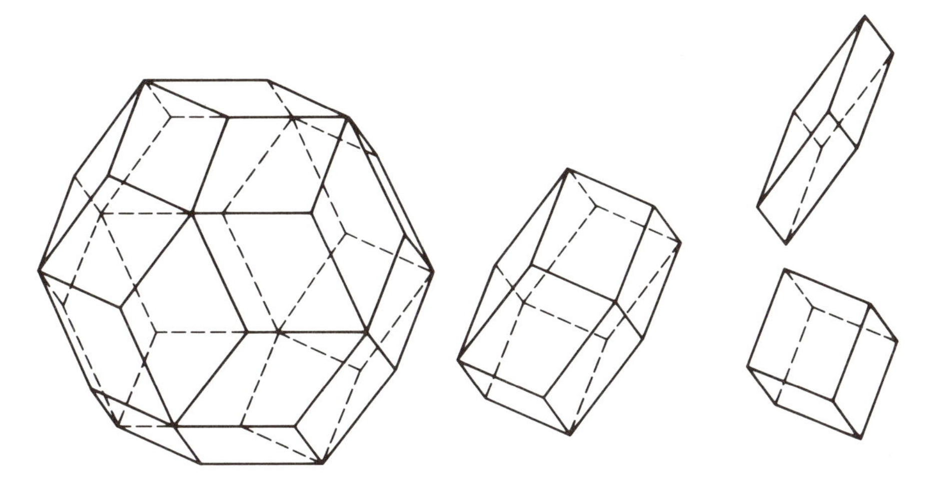

Figure 3.19 Skinny and fat rhombohedrons (*right*) can be fitted together to produce a triacontrahedron (*left*) or a dodecahedron (*middle*). The trick is to find a set of matching rules that guarantees that a set of basic building units can be fitted together to create a three-dimensional pattern with an icosahedral symmetry.

Such a complicated system is both cumbersome and inelegant. It's also hard to believe that atoms settling into an orderly atomic structure could readily sort out all the requirements for creating a perfect quasicrystal. However, no one has yet proved that this particular set of matching rules represents the minimum number required to build up a quasiperiodic structure having an icosahedral symmetry. Researchers have many more possibilities to investigate.

What all this has to do with real crystal growth isn't clear. In both the two-dimensional and three-dimensional cases, the models assume that atoms are already arranged in units corresponding to the fat and skinny diamonds or rhombohedra. But in a cooling liquid, individual atoms must come together first. There are no tiny tiles or polyhedra to provide the basis for tiling patterns—just the forces between different atoms.

Spheres in a Suitcase

Unlike cubes, identical spheres can't be packed together without wasting space. Cubes can fit neatly into a box, stacking in orderly layers to fill the space inside. A box filled with identical balls inevitably contains gaps between the balls. But how should the balls be arranged so that the maximum possible number fit within the box? In more general terms, what's the densest way to arrange identical spheres in space?

Such questions interest not only mathematicians but scientists as well. Random sphere packings play a role in describing and explaining the properties of liquids and granular materials. The study of sphere packings in higher dimensions leads to ways of designing digitally encoded messages to minimize power loss and confusion during transmission.

Everyday experience can mislead one into thinking the problem is trivial. Clearly, packing spheres in a cubic array, so that one ball sits on top of another, is less efficient than placing balls so they fit in the indentations between neighboring balls. In fact, shippers often use such a scheme for efficiently packing fruits such as oranges. Just arrange three spheres on a flat surface so they form an equilateral triangle. To complete the bottom layer, continue adding spheres so that each new sphere touches at least two spheres already in place. Build the second layer of spheres by placing each new sphere in the deep depression left at the center of any triangular group of spheres in the first layer. The finished second layer is identical to the first layer but slightly shifted horizontally.

Adding more layers by following the same rules results in a sphere packing called the face-centered cubic packing (see Figure 3.20). Such a packing fills just a little more than 74 percent of the available volume. Seen in fruit stands and in piles of cannonballs at war memorials, this arrangement is thought to be the densest packing that can be achieved.

But despite centuries of effort, mathematicians have not yet been able to prove that this is, indeed, the densest possible packing of identical spheres in three dimensions. It's conceivable that

Figure 3.20 The face-centered cubic packing of spheres is thought to be the densest packing of spheres in three dimensions. In the first layer of such an arrangement, the touching spheres form groups of triangles (*top*). Spheres in the second layer fit in the holes formed by a triangle of spheres in the layer beneath (*bottom*).

some irregular packing might be still denser. The best mathematicians have been able to do is to prove that no packing of spheres can occupy more than 78 percent of the space available. But such a limit doesn't help anyone looking for the most efficient way of packing spheres.

The sphere-packing problem, so simple to state and so difficult to solve mathematically, suggests that our mathematical understanding of three-dimensional euclidean space is not as complete as we would like to think. Most mathematicians and nearly all scientists believe the face-centered cubic packing is optimal.

They think nothing better will be found. But the path to a proof is strewn with obstacles.

Some of the complexities involved can be appreciated by following the logic of one putative attempt to find an optimal sphere packing. In three dimensions, the largest number of spheres that can be brought together so that each of them touches all its neighbors is four. The centers of these four touching spheres form the vertices of a regular tetrahedron, or triangular pyramid (see Figure 3.21). Because the four spheres can't move any closer together, this tetrahedral arrangement represents the densest possible configuration of four spheres in space.

The idea is to add new spheres to this tetrahedral arrangement one at a time so as to form a new tetrahedral pattern when-

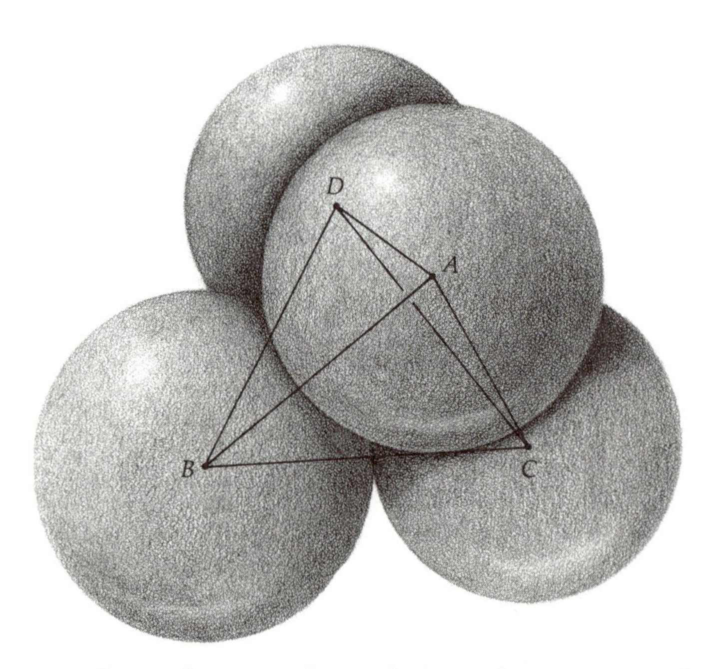

Figure 3.21 Four spheres can be packed together into a tetrahedral arrangement, but this pattern can't be repeated to fill all of three-dimensional space without leaving unfillable gaps. The resulting packing is less dense than the face-centered cubic packing.

ever possible. If such a procedure could be continued indefinitely, it would result in an arrangement of spheres with the densest possible packing. In fact, the proof establishing the presently known upper limit on the amount of space that identical spheres can occupy is based on just such an argument. Four nonoverlapping spheres fill about 78 percent of a tetrahedron's volume.

Unfortunately, tetrahedra don't fit together snugly to fill all of space. Cracks inevitably appear. Very quickly, the tetrahedral packing strategy leads to a bad move, and the exposed surface can't take a new sphere without wasting interior space. In the end, faithfully following the tetrahedral strategy as best as one can leads to a packing that overall is less dense than the face-centered cubic packing.

Mathematicians can generalize to any dimension the concept of a sphere and the problem of packing spheres. In one dimension, the "sphere" is just a line segment of unit length. Such "spheres," placed one after another in a row, cover 100 percent of an infinitely long, one-dimensional line. This packing is as dense as any packing can be. In the plane, the "spheres" are circles. The most dense packing of circles is a hexagonal lattice packing in which each circle is surrounded by six others (see Figure 3.22). The circles in this arrangement cover almost 91 percent of the surface.

Although a four-dimensional sphere can't easily be visualized, it's possible to define one analytically by analogy with the definitions for a circle in two dimensions and a sphere in three dimensions. Just as a circle is made up of all points equidistant from the circle's center, points on a four-dimensional sphere's "surface" are a fixed "distance" from its center (see Figure 3.23). Analogous definitions hold for higher-dimensional spheres.

Keeping track of spheres and their locations in higher dimensions is tricky. Packings in which spheres are irregularly or randomly arranged are especially hard to study, and computing densities or even specifying where centers of spheres are located is difficult. For this reason, mathematicians tend to stick to lattice packings in which spheres sit in a regular arrangement. The search for dense sphere packings focuses largely on lattice packings with highly regular configurations.

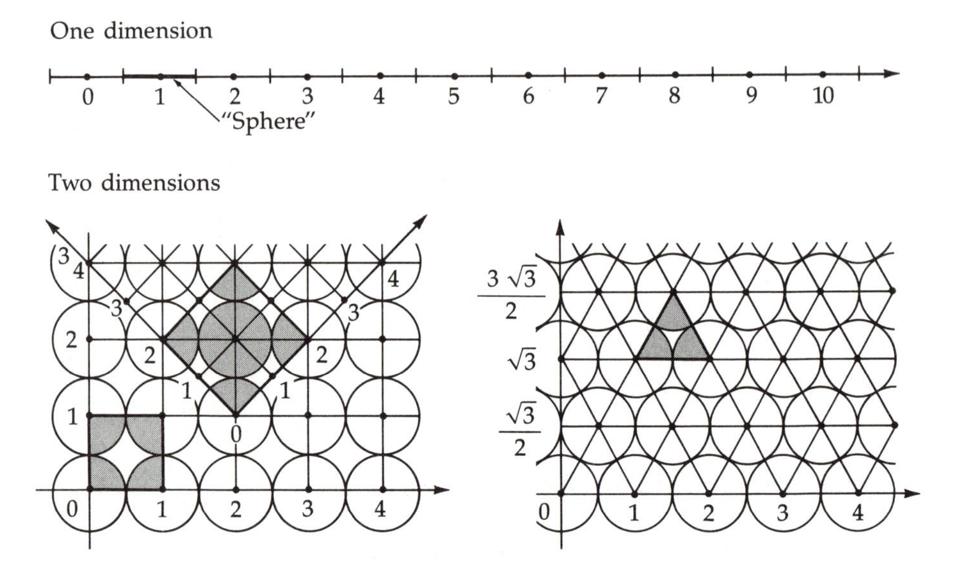

Figure 3.22 "Sphere" packings can be carried out in one and two dimensions as well as in three. In one dimension, the spheres are line segments of unit length. The spheres cover 100 percent of the line, and each sphere touches two others (*top*). In the plane, the spheres are circles, and there are three packings of interest. Two are equivalent square packings (*bottom left*), and the other, the densest packing of circles in two dimensions, is the hexagonal lattice packing (*bottom right*).

Mathematicians know that the densest lattice packing in two dimensions is the hexagonal lattice. Although it's possible that more dense, irregular sphere packings exist in three dimensions, Carl Friedrich Gauss proved in 1831 that the face-centered cubic lattice is the densest three-dimensional lattice packing. It's also known that analogous arrangements produce the densest lattice packings in four and five dimensions.

For dimensions higher than five, the appropriate analog of the face-centered cubic arrangement is no longer the densest lattice packing, and by the time one reaches eight dimensions, the gaps between the spheres are so large that it's possible to slide another copy of the same lattice into the available gaps without overlapping the spheres. In 1934, H. F. Blichfeldt proved that the lattice resulting from such a merger is the densest lattice in eight dimen-

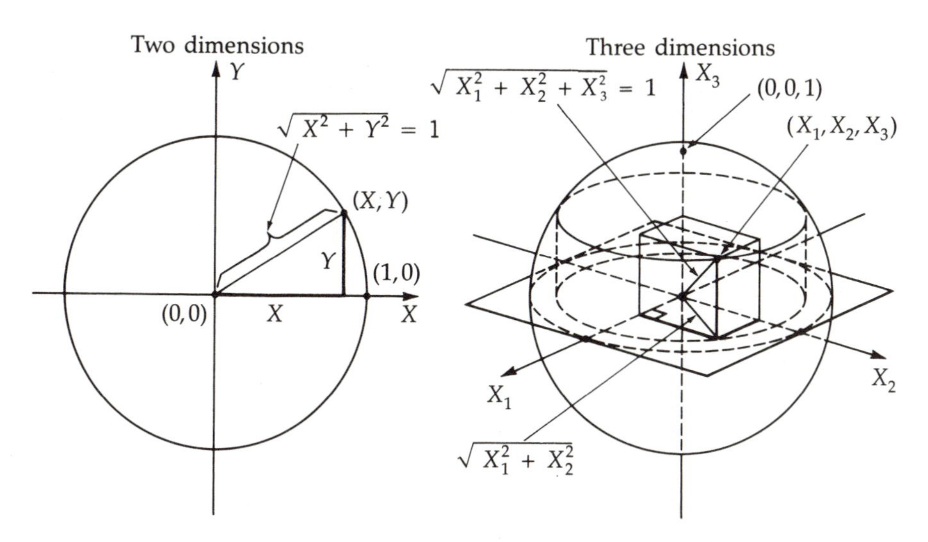

Figure 3.23 A point in two dimensions is specified by assigning a value to two coordinates, x and y. A circle of radius 1 around the origin (0,0) is the set of all points (x,y) that satisfy the equation $x^2 + y^2 = 1$ (*left*). In three dimensions, three coordinates, x_1, x_2, and x_3, are needed to specify a point. The surface of a sphere of radius 1 around the origin (0,0,0) is the set of all points (x_1,x_2,x_3) that satisfy the equation $x_1{}^2 + x_2{}^2 + x_3{}^2 = 1$.

sions. Moreover, he demonstrated that certain cross sections of this particular eight-dimensional lattice are the densest lattice packings in six and seven dimensions.

In 1965, John Leech constructed a remarkable sphere packing in 24-dimensional space. The study of the Leech lattice, as it is now called, has led to a deeper understanding of the properties of other higher-dimensional lattices and to important results in group theory. This particular lattice is almost certainly the densest possible sphere packing for a 24-dimensional space. Each sphere in the lattice touches 196,560 others. Very little is known about sphere packings in dimensions higher than 24.

It may come as a surprise to learn that the study of sphere packings in higher dimensions is more than just a game for mathematicians. The problem of packing spheres in n dimensions is mathematically equivalent to the problem of designing sets of digital messages that neither waste power nor cause confusion during transmission, especially over noisy lines.

In digital communication, the designer of a communications system is interested in constructing a list of distinct coded words that can be transmitted with maximum reliability and minimum power. Each code word may be represented as, say, an eight-digit symbol, each digit of which can take on one of five distinct values: 0, ½, 1, −½, or −1. At first glance, such a system provides 5^8, or 390,625, different words. However, the difference between many pairs of these words is minuscule and highly subject to electrical interference and to random errors that could easily change a transmission's meaning.

For example, a code word such as (1,1,1,1,1,1,1,1) is so much like the code word (1,1,1,1,1,1,1,½) that the two words could easily be confused. Stated in another way, the two words are so similar that a great deal of power would be needed to ensure that the words can be distinguished in the presence of background noise.

One way to design a signaling system that meets the necessary requirements is to represent each signal as a point in n-dimensional space. For example, the eight numbers of each code word in the signaling system described earlier can be plotted as a point in eight-dimensional space, where each number represents one of the coordinates. Thus, every code word in the system can be represented by a distinct point in eight-dimensional space.

First, for words to be reliably distinguishable, the points representing the possible code words must be separated by a certain minimum "distance." Second, the power required to transmit the words should be as low as possible. The total power necessary to transmit an eight-digit code word is the sum of the squares of all eight digits that make up the word. That sum is the square of the distance between a point representing the code word in eight-dimensional space and the origin, which is the point (0,0,0,0,0,0,0,0).

Thus, the design of a reliable, efficient signaling system is reduced to the geometric problem of placing points inside a region of space while keeping them from being too close together. If the points must be at least a certain distance apart, the problem is equivalent to the question of finding the densest packing of spheres whose radius is half that separation distance.

Suppose the eight-digit code words must be at least a distance $\sqrt{2}$ apart. Using the densest eight-dimensional lattice packing (discussed on pages 100–101), there are 240 points whose distance is $\sqrt{2}$ from the origin. Of those points, 112 have the form $(\pm1, \pm1,0,0,0,0,0,0)$, where the two 1s with any combination of signs can appear in any two positions, and 128 have the form $(\pm\frac{1}{2},\pm\frac{1}{2}, \pm\frac{1}{2},\pm\frac{1}{2},\pm\frac{1}{2},\pm\frac{1}{2}, \pm\frac{1}{2},\pm\frac{1}{2})$, where the number of minus signs is even. In this way, the densest eight-dimensional lattice packing could become the basis of a practical, efficient signaling system. If the scheme were to include exactly 240 code words, then the 240 points, each an equal distance from the origin, could be chosen as the code words.

Sphere-packing theory is just starting to be used for designing practical communications systems. It's the mathematics of efficient, reliable communication.

Computers in Kindergarten

A young child sits at a little kindergarten table that is covered by a grid of inscribed lines. Across this small field, the child sends armies of shapes—smooth, brightly colored cardboard squares, circles, and triangles, and maple-wood cubes, spheres, and tetrahedra. The child's fingers shift, stack, and arrange the forms to create pleasing, playful designs.

One child who went through this kind of experience was Frank Lloyd Wright. Later in life, as one of the most influential architects of the early twentieth century, he wrote: "On this simple unit-system ruled on the low table-top all these forms were combined by the child into imaginative pattern. Design was recreation! . . . The virtue of all this lay in the awakening of the child-mind to rhythmic structure in Nature—giving the child a sense of innate cause-and-effect otherwise far beyond child-comprehension."

Imagine a machine, instead of a child, playing at the kindergarten table. Suppose the machine studies this simple world of basic shapes and straightforward rules and begins altering this small world by creating artificial objects. If the machine makes something that we recognize as being playful and creative—something that we would expect a young child to make in such surroundings—then we could argue that the machine, too, is showing imagination. It may even be on its way to becoming an architect.

Imagination isn't necessarily a peculiarly human quality, says architect and mathematician Lionel March. "What I would like to do over the next twenty years is to build one of these machines and get it into a kindergarten,"he says. "It would look out on the world and make sense of it. It would have language and may have some theories to try out. It would act on this world and create things in it."

The idea of an "architecture machine" is not a new one. More than a decade ago, Nicholas Negroponte and his architecture machine group at MIT designed and studied sophisticated computer systems for intelligent computer-aided design. Their aim was to make machines that would create designs responsive to context and able to cope with missing information.

The MIT researchers decided that the important qualities required by an intelligent architecture machine include the ability to learn from experience and the capacity to make creative jumps. "Tools like intuition (sharpened by experience) are valuable and are often responsible for the major joys in architecture, and we should strive to bestow such devices on machines," Negroponte wrote in his book *The Architecture Machine*, adding, "My position is that machines, like humans, will have to evolve these mechanisms by developing in time and with experience, each machine being as different from the next as you are from me." Because public taste and needs also keep changing, the architect, whether human or machine, must be able to adapt.

In at least one sense, the concept of an architecture machine goes back centuries—to attempts by classical Greek and Roman designers, and later, Renaissance artists, to come up with "laws of beauty" and geometrical rules and relationships that would automatically translate into aesthetically pleasing structures.

The chief figure responsible for laying down the rules for art and architecture in fifteenth-century Italy was Leon Battista Alberti. An inventive man, attuned to the regular forms hidden within nature, Alberti based his rules on the geometrical concepts of proportion and ratio. Within his design for a building, for example, a certain proportion would be repeated over and over again on varying scales. Large arches would sweep over smaller but still perfectly proportioned arches, and the pattern would continue down to the finest details (see Figure 3.24).

For Alberti, design was governed by rules derived from a vocabulary of elements combined with a set of relationships be-

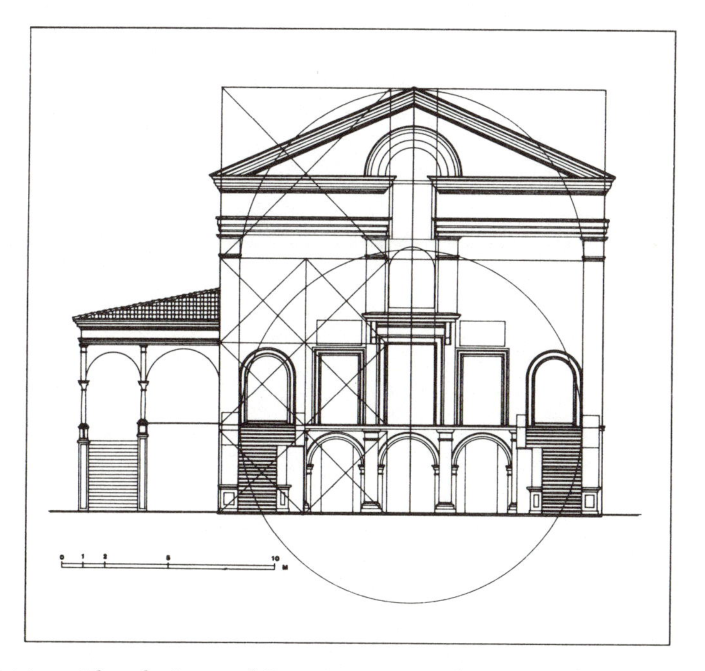

Figure 3.24 The designs of Renaissance architect Leon Battista Alberti reflect the careful use of ratio and proportion, as seen in his design for the facade of a church in Mantua, Italy.

tween these elements. This design structure provided a kind of "grammar" within which the designer worked. Later artists, including Leonardo da Vinci, incorporated and refined these ideas in their own works.

Renaissance architects, with a list of architectural features and procedures to choose from, produced a rich array of what many critics consider some of the most beautiful designs ever created. It was difficult to follow those particular procedures and not end up with an acceptable, pleasing design.

Architects today are pursuing, in their own way, grammars for design (see Figure 3.25). In particular, they are beginning to learn what mathematics has to offer, especially for delineating possible design choices. Applying modern mathematics to architecture is like shifting from the alchemy to the chemistry of design. New mathematical concepts—set theory, group theory and symmetry, graph theory and networks, mappings and transforma-

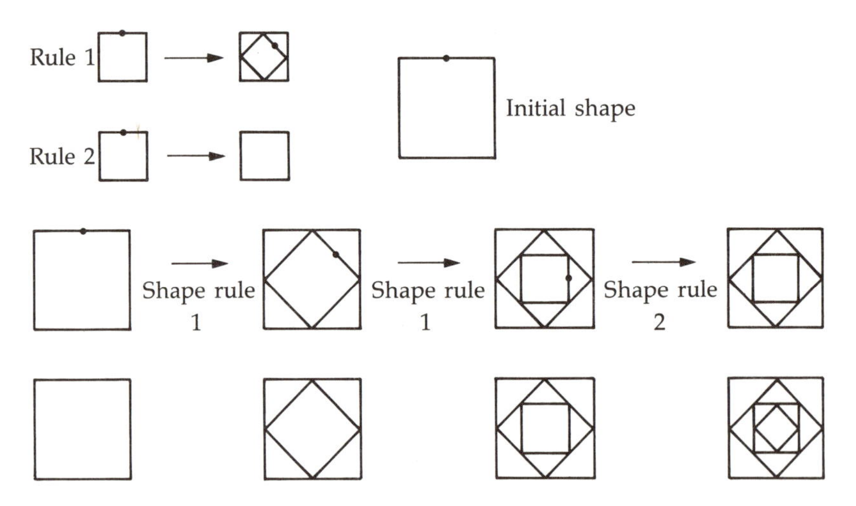

Figure 3.25 Shape grammars are algorithms that perform arithmetic calculations with shapes. Such calculations generate shapes with special structures and properties according to well-defined rules. In this simple shape grammar for embedding squares, two rules (*top left*) and an initial shape (*top right*) combine to produce various designs (*bottom row*) in the language defined by the grammar.

tions, and others—make it easier to describe and understand how shape is organized within buildings.

Mathematical arguments establish the existence of exactly 17 different kinds of repeating wallpaper patterns. Graph theory—the mathematical study of how points can be joined by lines—provides a useful way of representing floor plans when a designer wants to know how to arrange rooms and where to put doors so that rooms are connected in a reasonable way. Graph theory is also useful for arranging hallways in a large building and for planning the pattern of streets in a city.

Mathematical study also enables architects to know, in a precise way, how many fundamental floor plans exist. It's impossible to conceive of a building that doesn't include one of these basic plans. Although there are a limited number of such plans, architects still have considerable freedom to add their own distinctive touches. Take the example of three Frank Lloyd Wright houses that appear remarkably unlike in geometry but turn out to have the same basic pattern (see Figure 3.26).

Wright's work also showed frequent use of designs that began with a single, fundamental unit, which was then repeated to generate larger entities, going from a house to a large apartment building to a complex of towers. His procedure was very mathematical, relying on symmetry and the concept of "cyclic groups" to produce the designs, but the resulting structures were breathtakingly different from one another.

If architects can use mathematical knowledge in the design process, then computers can too. March's goal is to incorporate this kind of systematic mathematical knowledge into an architecture machine.

While it is possible to see how useful and fruitful mathematical ideas can be in the design process and how the process may be mechanized, the other half of the creative equation—imagination—is much more elusive. How do humans modify their stored, observation-based representations of the world to create something that doesn't yet exist, and then, when possible, act on the world to make the dream come true?

Some philosophers contend that language leads to imagination by providing a shorthand way of reworking representations of

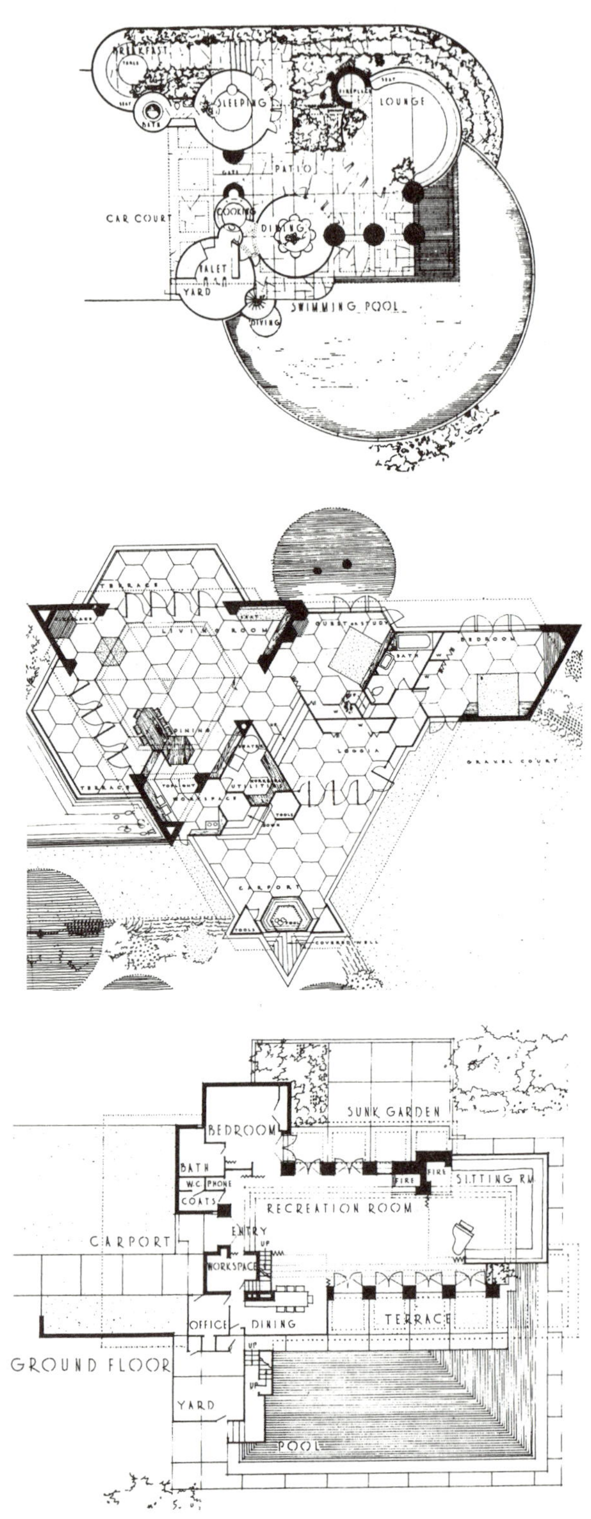

BREAKFAST
SLEEPING
LOUNGE
PATIO
CAR COURT
COOKING
DINING
VALET
YARD
SWIMMING POOL
DIVING
SUNK GARDEN
BEDROOM
BATH
WC
PHONE
COATS
FIRE
SITTING RM
RECREATION ROOM
ENTRY
UP
CARPORT
WORKSPACE
OFFICE
DINING
TERRACE
GROUND FLOOR
YARD
POOL

the world. Without language as an intermediary between the world sensed and the world effected, human behavior becomes simply a matter of stimulus and response—a matter of instinct.

With language as a vehicle, March and others see a possible way toward building a machine that would not only delineate choices but also show imagination. By providing a machine with a language for design, it could play with the rules to create new designs, just as Renaissance architects could work within the rules formulated by Alberti. Provided with an appropriate grammar, an architecture machine might attain the semblance of an artist's imagination.

Architects with an interest in mathematics have already attempted to formulate sample architectural grammars by studying in detail the work of particular architects. For example, a few years ago, two architecture students, working with George Stiny at UCLA, managed to extract the essence of Frank Lloyd Wright's work, and their computer generated new Wright house designs that experts insisted were authentic.

Of course, an architecture machine alone wouldn't handle the entire design process. The machine's ability would be based on a human language that someone would have to implant into the machine initially. Future architects may take on a new role as language designers, leaving the machines to figure out how to implement a certain architectural design.

By trying to mechanize the design process in this way, March believes that we can actually begin to understand more and more about what we do as human beings. The prospects of achieving artificial imagination are still remote, but efforts to create a science of design, based on mathematical principles, are advancing rapidly. Modern mathematical concepts now being developed

Figure 3.26 Sometimes, objects that appear to be very dissimilar may actually share an underlying structural pattern. These three houses designed by architect Frank Lloyd Wright, though they look completely different, are in fact topologically equivalent in plan. As shown by a simple application of graph theory, all three houses have rooms connected in the same way.

may help bridge the gap between the arts and the sciences and between human and machine imagination. A mathematical language of structure already has great value in biology and crystallography—in untangling the possible configurations of protein molecules and in establishing the structures of crystals.

4

Snowflake Curves

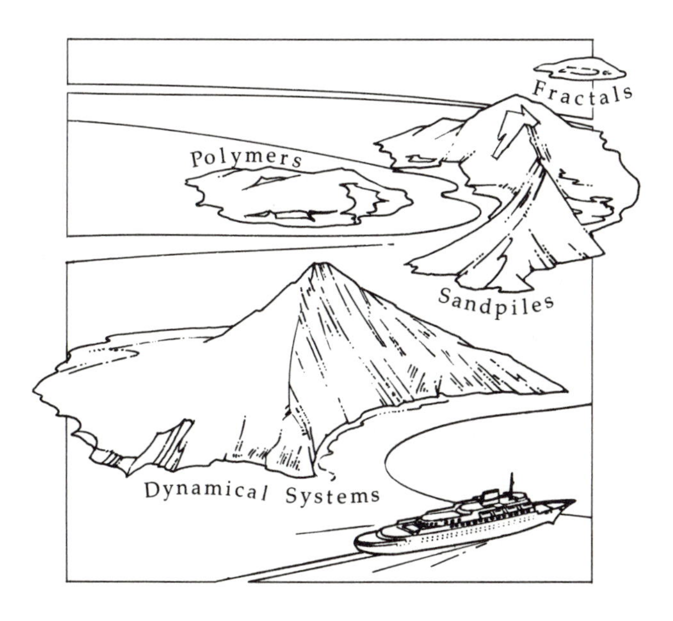

The branched patterns glistening on a frosted window, the jagged faces of a fractured rock, craggy mountains, billowing clouds, and rugged coastlines are difficult to describe in terms of triangles, circles, straight lines, cubes, and cones. Classical geometry fails to capture the kind of complexity seen in these and many other natural shapes and patterns. The trouble is that ordinary geometric figures lose their structure when magnified. Sufficiently enlarged, a portion of a circle begins to look like a featureless straight line; a sphere's rounded surface appears flat; a jagged curve starts to look smooth.

But there's more to geometry than triangles, circles, and spheres. It's possible to imagine mathematical curves and shapes that look just as complicated when they're magnified as they do in their original form. A closer look at a piece of such a figure reveals a miniature copy of the entire figure, and further magnification unveils even smaller copies. No amount of enlargement smooths away the structure (see Figure 4.1). Sometimes the relation of the part to the whole can be quite complicated. Like an image seen in a trick mirror, the magnified picture may look like a distorted version of the entire form, but a basic resemblance remains.

Such self-similar patterns are now known as *fractals*. Fractals add a rich array of dramatic shapes to the vocabulary of geometry. As a new language of nature, fractals provide both a description and a mathematical model for many of the seemingly complex forms found in nature. Self-similarity crops up as twigs sprouting from branches on trees, in the clumping of particles into dust balls, and in the system of veins and capillaries in the human body.

Fractal geometry allows researchers to restate old problems in fresh ways and to retell complicated stories more simply and comprehensively. The very existence of fractal forms suggests an enormous range of new physical and mathematical questions that take us away from the regular and the linear—away from

Figure 4.1 Enlarging a small portion of this fractal island's shoreline would reveal an array of tiny inlets duplicating the shoreline's overall pattern.

the smooth arc of a golfball sailing through the air and toward the tortuous path followed by a stream of smoke. Moreover, they bring geometry into problems where geometry doesn't seem to belong, providing a way of making sense of erratic behavior in apparently unrelated phenomena, such as soot formation, the stretching of silk fibers, and avalanches in sandpiles. Fractal geometry supplies new, exciting shapes for creating dazzling images of artificial but realistic worlds.

Fractal Forgeries

To draw a picture, a computer needs a set of instructions on how to proceed step by step. If an algebraic formula is available, the computer simply calculates the coordinates of points and then marks the appropriate ones on the screen. That's easy when the figure is a parabola. The computer uses the formula $y = x^2$, selects a value for x, computes y, plots the point, and goes on to the next value of x.

But what's the formula for a landscape of clouds and mountains? A simple landscape in which hills and clouds follow mathematical curves such as circles, ellipses, and sine waves looks quite unrealistic. Such a scene seems unnaturally smooth. But by using a mesh of tiny polygons, a computer can generate reasonably realistic graphic images. Unfortunately, that leaves the illustrator with the time-consuming and tedious task of specifying each one of the thousands of polygons that make up a single scene before a computer can conjure up the image.

A fractal description, on the other hand, provides a remarkably compact way of encoding the structure of a complex object. Because computer-generated fractal images have similar patterns on many different scales, computer programs for creating even intricate images are relatively short. Software written to produce the detail on one scale can be slightly modified and reused in a loop to repeat the image on successively smaller or larger scales. Indeed, the simple, repeated operations that go into the construction of a fractal are ideally suited to the way a computer functions. The computer patiently performs the same set of operations over and over again to generate the fractal object.

The construction of one intricate fractal form, called the snowflake curve, starts with a single line segment divided into thirds. Replacing the middle portion by two equal segments to form two legs of an equilateral triangle puts a sharp jog into the original line. The next stage involves dividing each of the four segments in the new figure into thirds. Replacing the middle third in each of these smaller segments by two equal pieces brings more jogs to the curve, and so on. Each step, or iteration, adds more detail. The eventual result is an infinitely crinkly curve (see Figure 4.2).

Like all fractals, the snowflake curve has unusual geometric properties. Although the recipe for generating the snowflake curve is concise, simple to describe, and easily computed, no single algebraic formula specifies the curve's points. Moreover, at each stage in its construction, the snowflake curve's length increases by a factor of four-thirds. In the end, it crams an infinite length into a finite area of the plane, yet doesn't intersect itself.

Of course, the fractal image that appears on a computer's video display isn't really, in a mathematical sense, a true fractal. A computer generating a snowflake curve, for instance, can display no line segments smaller than the tiny spots of light, or pixels, dotting the screen. The true snowflake curve consists of even smaller line segments. Computer graphics necessarily provides only an approximate, though useful, picture of fractal forms.

In everyday geometry, a line fills one dimension, a square two dimensions, and a cube three dimensions. Fractals usually have fractional dimensions. A one-dimensional line that curves in and out within a given area may eventually fill up the whole two-dimensional area. Such a pattern has a fractal dimension of 2 (see Figure 4.3). In contrast, the snowflake curve, though extremely convoluted, fills less of the plane, and for this reason, has a smaller fractal dimension, which happens to be approximately 1.26.

The notion of fractal dimension provides a numerical measure of the wiggliness of a curve or the roughness of a surface. It fits within the intuitive idea that a crinkly curve somehow covers more of a given area than a smooth curve and that a rough surface fills more space than a smooth one. Beyond that, a fractal dimension is a way of expressing the fact that a fractal is self-similar, or contains copies of itself. It reflects the relationship between the whole and its parts.

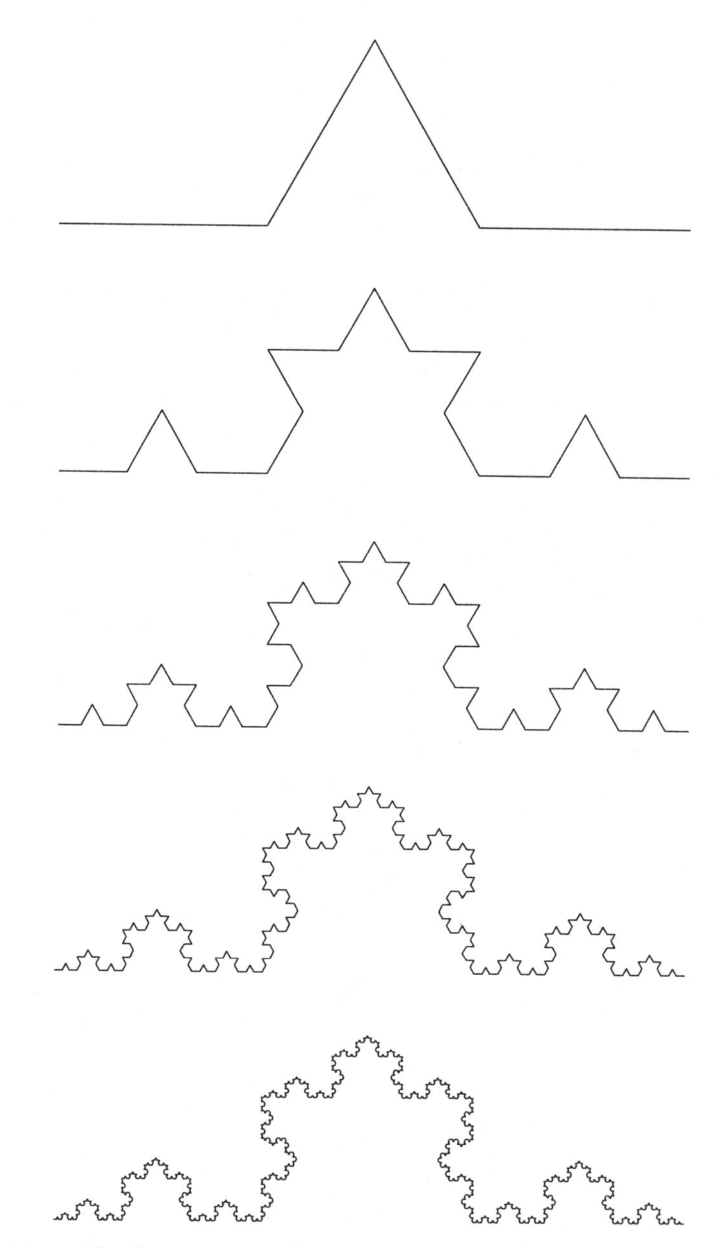

Figure 4.2 The first six stages for generating an infinitely crinkly snowflake curve.

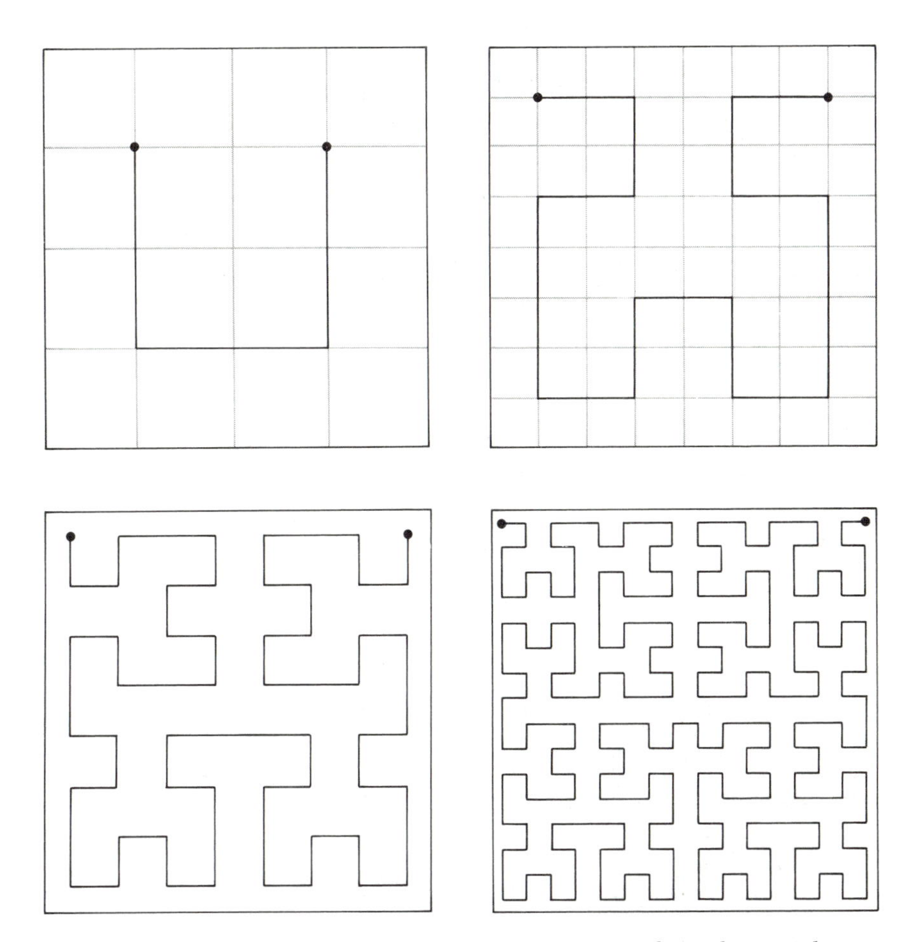

Figure 4.3 The first four steps in a recursive procedure that produces a space-filling curve. In the end, the curve becomes a solid square.

It's possible to create fractal cousins of the snowflake curve having any dimension between 1 and 2. Although these curves have a fractional fractal dimension, they are still, in another sense, one-dimensional curves. Removal of a single point would cut the curve into two pieces.

A smooth object with a smooth surface, such as a flat sheet of paper, is clearly two-dimensional. Crumple the paper into a ball, and the result is a rough surface having a fractal dimension lying somewhere between 2 (the dimension of a plane) and 3 (the dimension of a solid sphere). In a three-dimensional setting, a fractal object with a dimension of 2.9 would look like a sponge, whereas one with a dimension of 2.3 would resemble a randomly pock-marked golfball (see Figure 4.4).

In computer graphics, tinkering with the fractal dimension dramatically alters a scene. As the fractal dimension rises, flat plains and gently rolling hills swell into towering mountains, which grow jagged peaks. When the fractal dimension falls, the peaks gradually wear away, and the mountains retreat into a gentle landscape.

Whereas fractal landscapes show similar features on all scales, natural landscapes are self-similar only over a limited range of distances. The largest possible variations in a mountain's profile are limited by the force of gravity, and the smallest varia-

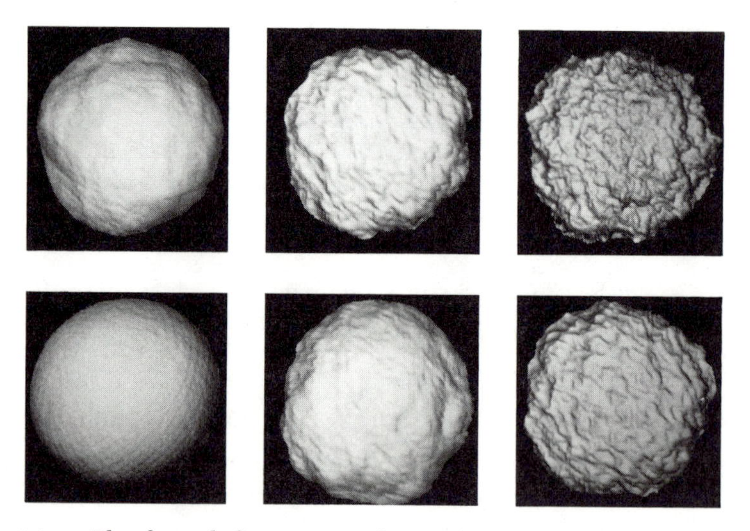

Figure 4.4 The fractal dimension of a surface depends on the ratio between the altitude of the larger bumps to that of the smaller bumps. These spherical shapes have surfaces with fractal dimensions ranging between 2.0 (smoothest) and 2.5 (roughest).

tions are governed by the smoothing effect of erosion, the basic size of grains in rock and sand, or, at the very least, the finite size of atoms. Fractal geometry provides only a convenient mathematical approximation to the real world.

Fractal valleys don't look nearly as realistic as fractal mountains. One simple way to overcome that problem is to fill the basins to a certain level with water, hiding the less realistic low points while retaining the majesty of the mountain peaks. Because fractal geometry specifies only relative height variations at different length scales, illustrators have also tried to solve the problem by mathematically manipulating these variations: cubing the height variations on a ragged, fractal curve will flatten the lower elevations of a mountain landscape and keep the higher elevations rough. Alternatively, taking the cube root of height variations achieves the impression of river erosion in an otherwise fairly smooth plain (see Figure 4.5).

In the end, most of the time spent in producing a fractal landscape image is not in the mathematical work of computing heights above a base plane but in shading and coloring the picture to make it look convincing. Simulating pearly highlights, the bleeding of color from one surface to another, and subtle gradations of reflected light has forced researchers to take a closer look at how light behaves as it reflects from surface to surface, whether textured or mirrorlike.

Drawing a realistic tree is one of the major unsolved problems in computer graphics. Drawing a fractal, branched structure is fairly easy. Branches, subbranches, and twigs all repeat the shape of the tree as a whole on different scales. The routine continues until the smallest twig is a specified fraction of the size of the main trunk, at which point the computer program shuts itself off (see Figure 4.6). But fractal trees drawn in this way always seem more sparse than real trees.

One drawback of simple fractal models is that they fail to reflect changes in the branching pattern of real trees as the scale changes. For example, pine needles don't have exactly the same shape as the branches to which they're attached, nor do the branches replicate the shape of the tree as a whole. Moreover, real

Figure 4.5 Valleys in mountain landscapes based on fractal curves rarely appear as smooth as they do in natural settings. One answer is to hide the "unsmooth" valley by covering it with a lake. One can also alter the vertical scale, for example, by raising the altitudes above the valley bottom to the third power (*top*). In contrast, by raising the altitudes to a power less than 1, one obtains steep-walled mesas and canyons (*bottom*).

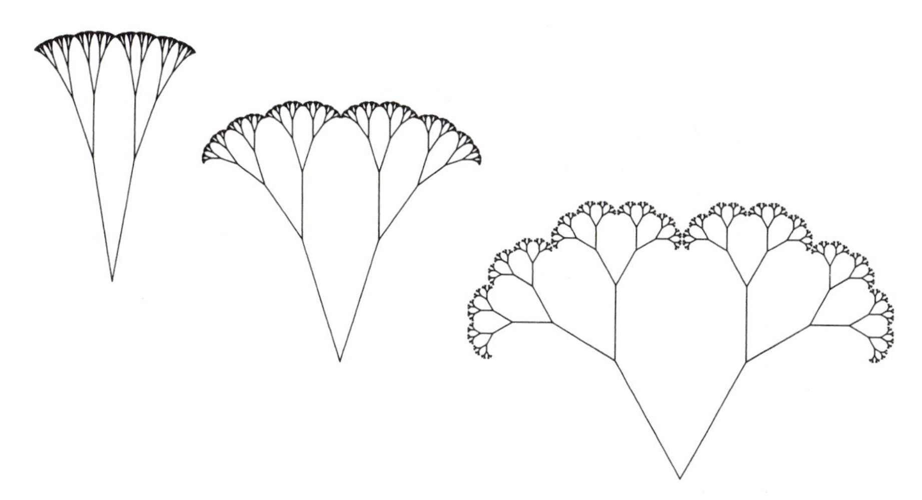

Figure 4.6 A branching process in which each branch splits into two shorter branches generates a fractal canopy.

trees grow so that their branches and twigs don't overlap. A fractal approach doesn't allow for this kind of feedback.

One promising approach is to combine the branching character of a fractal structure with just the right dose of knowledge about how plants grow (see Figure 4.7). It turns out that surprisingly crude assumptions about plant growth lead to remarkably realistic pictures. A formula, with small random perturbations, sets the angle and position of each set of branches. The resulting figure looks much like a natural tree. More sophisticated programs for drawing trees include additional parameters such as the curvature and helical twist of the branches.

One of the most remarkable ways of creating complex fractal images combines a set of very simple rules with randomness in the form of tossing a coin or throwing a die for deciding which rule to use at any given moment. For example, start with a triangle drawn on a piece of paper. Label the triangle's corners 1, 2, and 3. Mark a starting point anywhere within the triangle. Three rules specify where subsequent points fall. The roll of a die or some kind of random number generator specifies which rule to use. If

Figure 4.7 Adding some rules about the way plants grow produces more realistic trees.

the number 1 comes up, the new point falls halfway between the previously specified point and the triangle vertex labeled 1. If the number 2 comes up, the new point falls halfway between the previous point and vertex 2. If 3 comes up, the new point falls halfway between the previous point and vertex 3.

At first, the points seem randomly distributed, but after 100,000 or so turns, a definite pattern begins to emerge. In this particular example, the figure turns out to be not a randomly distributed dust of points but a highly regular fractal known as the Sierpiński gasket, which consists of triangles nested within triangles—a triangular plate punctured by an infinite number of triangular holes (see Figure 4.8).

Different sets of rules produce different fractals. With enough rules, a computer can generate not only fractal curiosities but also

Figure 4.8 The Sierpiński triangle, or gasket, emerges from a million or more random choices made among three rules that, in effect, transform space.

convincing images of natural objects such as leaves, ferns, clouds, and forests (see Figure 4.9).

This dynamic approach to drawing fractals is based on ideas first proposed in 1981 by John Hutchinson and later developed by Michael Barnsley. The rules are really mathematical operations known as *affine transformations.* An affine transformation behaves somewhat like a drafting machine that takes in a drawing, then shrinks, enlarges, rotates, or skews the picture, and finally spews out a distorted version of the original. In other words, the transformation "maps" each point in the plane to a new location. The most important mappings for drawing fractals are those that contract the plane by moving points closer together.

Figure 4.9 Small adjustments in the rules for generating a fern lead to small changes in the resulting shapes.

Any set of contractive transformations, or mappings, produces a fractal picture. The procedure can be further fine-tuned by adjusting the probability with which any particular mapping is chosen. By keeping track of how often each pixel is visited, the computer can also assign colors or shades of gray to the resulting fractal form (see Color Plate 9).

In fractal forms, researchers and artists working with computers have a sophisticated, responsive pallette for producing both lifelike images and intricate abstractions. Using simple rules repeatedly applied, they can generate startlingly realistic yet eerily alien fractal landscapes featuring randomly cratered surfaces, or craggy mountains towering over lakes with erratically indented shorelines.

Fractal images show that simple rules can lead to complicated structures and remind us that nature's diversity may have roots in similarly simple fundamental principles. Fractals also tell us how large-scale form is related to small-scale detail. They remind us to look at objects on all scales, because there are interesting features to be seen wherever you look.

Written in the Sky

One of the marvels of nature is the way simple water molecules can settle into symmetrical structures having the seemingly unlimited variety and intricacy of snowflakes. Little is known about how these lacy shapes are created. Why does a snowflake branch? How does one branch tell another which way it's going so that the whole flake stays more or less symmetrical?

For centuries, snowflakes have attracted attention for their striking blend of symmetry and intricacy (see Figure 4.10). More than 2,000 years ago, Chinese scholars recognized and noted the characteristic six-sided or six-branched form of snowflakes. In 1611, astronomer Johannes Kepler wrote a little book in which he speculated about why snowflakes always fall as six-pointed starlets. He pondered hexagonal packings of spheres at a time when the existence of atoms was little more than speculation, but he couldn't solve the problem of snowflake formation.

Early investigators, lacking the sophisticated tools and high level of mathematics needed to solve the mystery, had to be content with merely classifying snowflake patterns. These keen observers of nature couldn't help noticing that despite similarities in overall shape, no two snowflakes ever appear identical in every detail.

Scientists are only now beginning to decipher the message contained in a snowflake's pattern and to form a theory of how snowflakes grow. They see an intense, exquisitely balanced competition that influences a snowflake's final form and explains why snowflakes come in such endless variety. Indeed, a snowflake's delicate branches record a remarkable history of a tumultuous journey through the air. Each feathery feature is a result of subtle variations in temperature and humidity.

Fundamentally, snowflakes are formed in a purely mechanical way. The basic building blocks are water molecules. It takes many billions of billions of water molecules, each one settling into the right spot, to create even a tiny snowflake that is barely visible to the naked eye.

Figure 4.10 The characteristic hexagonal symmetry of snowflakes has long intrigued scientists and scholars.

One secret is revealed by comparing snowflake formation with the freezing of water in a pond or in a refrigerated ice tray. The water doesn't normally freeze into a branched pattern. Ice first forms at the container's walls, then gradually spreads smoothly toward the middle. The walls drain away excess heat, which represents the energy that water molecules give up when they stop moving and settle into place. Snowflakes, however, freeze and take shape in moist air, free from any walls. A typical snowflake begins as a dust particle or some other airborne impurity. That particle snares some of the water molecules that happen to be wandering about nearby. Gradually, as more molecules arrive, a microscopic layer of ice forms.

As it takes on water molecules, the snowflake must get rid of its excess heat to keep growing. That happens most efficiently when the snowflake has a wrinkled, rather than a smooth, surface. Because added roughness increases its surface area, the more it becomes like a pincushion rather than a ball, the more effectively a burgeoning snowflake can shed heat.

How quickly and readily heat diffuses is governed by how steeply the temperature changes near the snowflake's surface. The steeper the temperature gradient, the faster snowflake growth will be at a given point. But that process is complicated by the fact that settling water molecules themselves release heat, warming the neighborhood. That heat must be removed before further solidification can take place.

In principle, physicists can use a mathematical expression known as the Laplace equation to describe how heat diffuses away from an object. Finding solutions of the diffusion equation is relatively easy for a warm metal ball or a smooth, heated pipe immersed in air or water at a constant temperature. But it's much more complicated for a snowflake, where the arrival of each water molecule changes the object's shape and the local temperature. Computing the heat flow would be like trying to play a tune on the piano if each note played determined each subsequent note. You'd have no clear idea of where you were going until you arrived.

To overcome the mathematical difficulties inherent in solving equations that essentially represent moving targets, researchers have taken two different approaches. One group emphasizes finding solutions of simplified, special cases of the equations to gain insights into which factors are most likely to affect snowflake formation. A second group uses computer simulations of the random aggregation of particles. They convert the difficult-to-solve equation governing heat diffusion into a sequence of simple steps repeated over and over again. Which approach works better is still a matter of controversy.

One of the simplest possible growth models is known as *diffusion-limited aggregation*. In this model, a cluster grows one particle at a time, as each wandering particle comes into contact with and sticks to the growing object. The surprising result is a branched, fractal structure (see Figure 4.11), even when the initial "seed" is smooth and featureless. Randomly wandering particles are more likely to stick at or near a bump or tip. If a particle happens to wander into a hole or channel, it's almost certain to stick to the hole's side before it gets to the bottom. Thus, departures from a regular shape grow rapidly, and protruding tips wander and split repeatedly to produce a fractal pattern. The resulting structures are typically wispy or tenuous.

Although the diffusion-limited-aggregation model does produce forms that resemble electrochemically deposited mossy zinc and the electrical discharge patterns created by large voltages applied to materials, such branched particle aggregates don't look like snowflakes. A typical snowflake is symmetrical. It stabilizes at some point, retaining its shape and continuing its growth in definite directions, although it may spawn side-branches as it grows. Without this stability, every snowflake would be an extremely wrinkled, feathery object. Growth would also be subject to the tiniest temperature shifts. A branch that happened to encounter a relatively cool region would shoot out faster than the others. Long branches would grow faster than short branches, and the result would look nothing like a real snowflake.

Such a pattern of unstable growth does, however, explain how tiny differences at the molecular level can be translated into a snowflake's characteristic six-armed pattern. When water molecules form into ice crystals, they settle into a hexagonal

Figure 4.11 In diffusion-limited aggregation, randomly wandering particles eventually come in contact with a growing aggregate, sticking and becoming part of the cluster. The result is a wispy, branched structure, even when the initial "seed" is smooth.

arrangement—a kind of three-dimensional honeycomb (see Figure 4.12). That structure provides a slight, built-in preference for growth in six directions. Once tips pointing in the six directions start to form, they grow faster than the rest of the crystal. The beautiful macroscopic structure of a snowflake reflects the underlying hexagonal lattice in which its molecules are arranged.

Figure 4.12 The faces of a snow crystal (white shapes) reflect the orientation of the underlying hexagonal lattice of ice molecules (dots). Some areas of the crystal's surface grow faster than others, giving rise to a great variety of crystal shapes having a hexagonal symmetry.

Surface tension—the tiny, almost negligible force that actually holds a snowflake together—is the stabilizing factor. It acts like a plastic skin that restrains the faster-growing branches. In effect, surface tension holds a tip back, allowing time for the rest of the object to catch up with it.

Mathematical analysis and experiments confirm the important but subtle role played by surface tension. Studies also show the importance of tiny anisotropies, or distortions, caused either by the crystal's geometric arrangement or variations in the temperature and humidity surrounding the growing snowflake. Without anisotropy and surface tension, the branched, or dendritic, growth of snowflakes couldn't occur.

Given the tremendous variations among different snowflakes, why do the six arms of a single snowflake look so much alike? Experimental studies reveal that environmental conditions at the six tips are a lot more similar than the environmental conditions between one snowflake and another. The six arms of a snowflake, each less than a millimeter long, tend to see the same temperature distribution and vapor density. Two neighboring but separately drifting snowflakes are far enough apart to encounter significant differences. However, the conditions surrounding even a single, tiny snowflake aren't completely uniform. A close look at any real snowflake reveals that its six arms aren't identical. There are always some imperfections.

Can a computer simulate snowflake growth? Physicist Fereydoon Family and his colleagues have demonstrated one scheme that produces convincing snowflake replicas. Researchers first draw up a honeycomb grid as large as their computer can handle. Each space, or cell, is a potential home for one water molecule. At the center sits the "seed"—a small number of marked cells corresponding to the dust particle at the heart of a snowflake. For each space around the seed, the computer calculates the temperature gradient at an instant in time. If the gradient is larger than a certain value, then a water molecule is allowed to occupy the space and the computer-generated snowflake grows by one unit. The same procedure is repeated over and over again, and the spaces gradually fill.

The resulting shapes depend strongly on the selection of the cutoff value determining whether or not a space is filled. Randomly changing the value at each step creates rough, irregular forms that look like soot particles. Keeping the value constant produces lacy, fractal shapes that look much too regular to be snowflakes (see Color Plate 10). However, choosing the value so that it varies smoothly in a certain way produces patterns that are very similar to real snowflakes (see Figure 4.13, top).

Figure 4.13 By varying the growth rules, researchers can generate a wide variety of branched, or dendritic, patterns similar to real snowflakes (*top*). Changing the rules as a single pattern grows creates a layered structure (*bottom*).

Real snowflakes also have a layered structure, reflecting the varying conditions that influence their growth as they toss and swirl on their way to earth. A snowflake's final coat often hides a range of different structures buried deep inside. It's possible to simulate this layered structure by regularly changing and manipulating the cutoff value (see Figure 4.13, bottom). Those manipulations correspond to changes in growth conditions at different times.

A falling snowflake carries the clues for deciphering many mysterious, complex patterns in nature. It's an extremely sensitive and selective amplifier of minute fluctuations in the environment. From studies of the layers in snowflake patterns, physicists may develop insights into the atmospheric conditions that a snowflake encounters on its turbulent journey to earth. That, in turn, could lead to a deeper understanding of cloud formation, storm development, and other atmospheric phenomena. And because branched patterns are so common in nature, snowflake studies could shed light on why lightning follows a jagged path through space, and why metals and other materials sometimes crystallize into treelike, branching forms.

Time to Relax

The rubber in a pair of boots, retrieved after a long stay in an attic, shows its age in an annoying way. No longer as flexible as it once was, the material (an elastomeric polymer) readily cracks and falls apart. Under the same conditions, many other plastics suffer a similar fate.

One cause of this aging process is chemical. Sunlight or oxygen can initiate chemical reactions that alter the material's properties. But deterioration occurs even when a material is kept in the dark or away from oxygen. The material gradually becomes more dense and brittle, losing its toughness and impact resistance.

The explanation for this behavior lies in the way "defects" within amorphous, or noncrystalline, materials reorganize themselves over long periods of time. When expressed in terms of the concept of fractal time, the same mathematical model used to describe polymer aging also applies to the stretching of glass or silk fibers; the recovery, or *relaxation*, of glassy materials after the removal of a stress; and a wide range of other phenomena in amorphous materials. In such processes, events occur in self-similar bursts—featuring distinct clusters of activity interspersed with long stretches of inactivity. Some changes in materials occur right away while others take years to show up.

Relaxation is an issue of practical importance. Slow aging processes, both environmental and physical, control the lifetimes of a great many manufactured products, from electronic devices to optical fibers and advanced composite materials. The new theory of how such processes occur suggests novel techniques for toughening ceramics and for designing polymers having particular characteristics.

Relaxation processes are common in physical systems. For example, pull on a glass fiber, then let go. The glass first stretches, then shrinks. Apply a strong electric field to a polymer, then turn it off. Areas of positive and negative charge in the polymer line up with the field, then drift out of alignment. In each case, the material endures a stress, then recovers, or relaxes, when the stress is removed.

Relaxation in a crystalline material typically proceeds at an exponential pace. That type of relaxation follows the same pattern as the decay of a radioactive isotope. Such a process is characterized by a certain time, known in the case of radioactive decay as the half-life.

Normally, one finds that relaxation is clustered around a certain time. It might take a second, a day, or a week. But physicists have discovered that an amorphous solid takes a longer time to relax than would be expected if the relaxation simply followed an exponential decay. In amorphous systems, some parts relax very quickly. If those parts relax in, say, seconds, other pieces might relax on a time scale of minutes, and still others on a scale of days or weeks. If researchers were to wait long enough—even years—they would still detect changes taking place. No characteristic time can be defined for such an extended relaxation process.

This type of behavior has come to be known as *stretched exponential relaxation*. It fits a wide range of relaxation processes in disordered systems, including the way many polymers, glasses, and ceramics respond to stresses caused by changes in pressure and temperature and the imposition of electric and magnetic fields.

Because so many different systems behave in such a strikingly similar fashion, physicists, in their search for an explanation, have concentrated on what these systems have in common. They have found that what's important is not the details of a material's atomic and molecular structure but rather its state of disorder.

An amorphous material's constituent atoms or molecules lie in random positions rather than at well-defined sites in an orderly crystal lattice (see Figure 4.14). Moreover, just as crystal structures are rarely perfect and contain dislocations, vacancies, and other imperfections, amorphous materials also contain "defects," in which bonds between atoms or molecules may be strained,

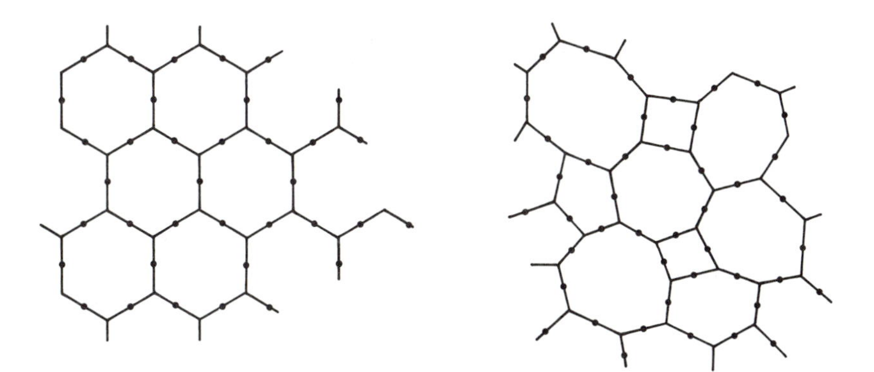

Figure 4.14 In a crystalline material, atoms or molecules sit in an orderly array (*left*). In amorphous solids, the pattern is much more irregular (*right*).

distorted, or displaced. For example, such defects occur during glass formation because molecules find they have too little time during the cooling process to orient themselves into their proper positions to form a closely packed crystal structure. Inevitably, glasses end up containing low-density regions, the analog of vacancies in crystals.

In 1983, physicists Michael Shlesinger and Elliott Montroll proposed that migration, or diffusion, of mobile defects could account for stretched exponential relaxation in the case of an amorphous material relaxing after the application of an electric field. They suggested that defects, in order to move, have to overcome different energy barriers scattered throughout the material. Whereas small barriers are easy to hurdle, larger ones significantly reduce mobility.

In the early stages of relaxation, defects that are near low barriers don't have any trouble. There's enough thermal energy for them to jump and cause relaxation. Others, faced with moderate barriers, take longer to get moving. Thus, a random distribution of energy barriers implies a wide range of relaxation times, leading to the stretched exponential relaxation observed for amorphous materials. Relaxation stretches out over a long period of time.

Mathematically, the situation is closely related to the problem of determining the length of a fractal. Magnifying a fractal by any amount reveals a miniature version of the larger form. Finer and finer scales show more and more detail and lead to greater and greater estimates of total length.

For example, measuring the length of a fractal coastline leads to different answers, depending on the scale used. On a world globe the size of a basketball, the eastern coast of the United States looks like a fairly smooth line, which, according to the globe's scale, may be roughly 3,000 miles long. The same coast drawn on an atlas page showing only the United States looks much more ragged. Adding in the lengths of capes and bays extends the coast's length to 5,000 or so miles. Piecing together detailed navigational charts to create a giant coastal map reveals an incredibly complex curve perhaps 12,000 miles long. Each change in scale reveals a new array of features to be included in the measurement.

Just as every distance scale occurs in the coastline problem, every time scale occurs for relaxation in amorphous materials. Each shift in time scale—from seconds to minutes to days to years—adds new features to be included in a relaxation measurement. Although it isn't as picturesque to think of infinitely many time scales as it is to think of patterns within patterns on different length scales, the analogy is mathematically exact.

This comparison leads to the concept of fractal time. Instead of occurring in a sequence of regular, equally spaced intervals, events that occur in fractal time are clustered. Such clusters consist of events that happen rapidly, one after the other, interspersed with long stretches of nothing happening in between (see Figure 4.15).

To support this theoretical picture, researchers have discovered that in polymer relaxation, some phenomena occur within picoseconds whereas other effects aren't apparent for years. Such an astonishing array of time scales shows how tricky it is to do experiments investigating the phenomenon because it's hard to measure things over so many orders of magnitude in time.

John Bendler and his colleagues at General Electric have applied the theory to understanding the properties of Lexan, a tough polycarbonate resin used for making bulletproof windows for limosines. Defect diffusion turns out to be a good model for how the material responds to stresses and how it ages.

A chunk of Lexan consists of an irregular, three-dimensional network of long polymer molecules, each one a twisted chain thousands of atoms long, with a precisely defined, repeating pat-

Figure 4.15 Instead of occurring at regular intervals, events that happen in fractal time are clustered in a self-similar pattern that features rapid bursts interspersed with long pauses.

tern of atoms. Experiments indicate that cooling the polycarbonate results in the freezing in of a small population of high-energy kinks in the molecular chains (see Figure 4.16).

It's the movement of these kinks in a fractal-time process along the polycarbonate chains that leads to relaxation. Kink movements reorganize the molecular backbone and effectively absorb mechanical energy, such as the impact of a bullet or a

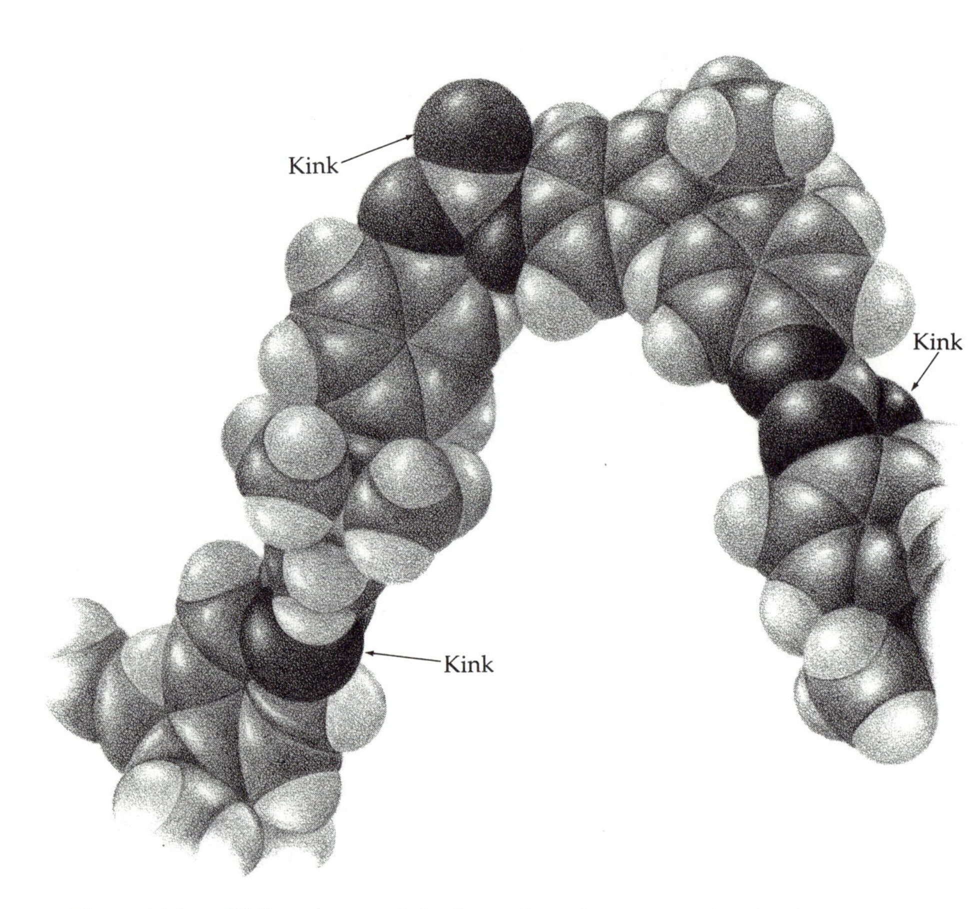

Figure 4.16 Molecular model of a polycarbonate molecule showing the carbonate linkages that act as kinks. The fractal-time motion of such kinks leads to stress relaxation.

sledgehammer. They are also responsible for aging. As energy-absorbing kinks reach chain ends, the material gradually becomes more brittle and weaker. With that insight, researchers now believe that it may be possible to slow aging by modifying chain ends.

The fractal-time, or defect-diffusion, model also helps to explain the stretching of silk and glass threads. In 1835, German physicist Wilhelm Weber noticed that attaching a weight to a long thread causes it to stretch to a certain length immediately. But that instantaneous elongation is unexpectedly followed by a gradual further lengthening that depends on how long the weight is applied.

The reason for such behavior lies in the fractal-time movement of defects within the materials. Silk is a complicated natural polymer and has a variety of different amorphous and crystalline forms. Under an applied load, the material tries to rearrange itself to redistribute and minimize stresses. Under those conditions, silk molecules relax by unwinding and changing the hydrogen bonding along their backbones. In a glass fiber, the mobile defects correspond to imperfections in the distorted, tetrahedral network of oxygen and silicon atoms. Under a load, materials such as silk and glass mechanically reorganize themselves.

Although ceramicists, engineers, and artisans such as glassblowers have long been aware of the peculiar behavior of glasses, polymers, and ceramics and have taken these properties into account when working with the materials, little progress in understanding relaxation phenomena occurred until recently because the mathematics initially used to describe such processes seemed so complicated and difficult. The new concepts of mobile defects and fractal-time motion appear to provide a more tractable, self-consistent picture of the relaxation behavior of supercooled liquids and glassy solids.

One of the chief merits of the new theory is that it's mathematically simple. Researchers can use defect-diffusion mathematics — the mathematics of intermittent pausing — to model the kind of behavior displayed by almost all amorphous materials. The theory also suggests ways of modifying in a useful manner the properties of industrially important materials.

Life on the Edge

One of the puzzles of modern physics is why fractal geometry seems appropriate for describing such a wide range of physical phenomena, from mountain landscapes, coastlines, and the distribution of galaxies to turbulence, the luminosity of stars, and noise in electronic devices. In each instance, the phenomenon shows a fractal structure in time or in space. It occurs on all measurable distance or time scales. There's no characteristic length or time.

In 1987, Per Bak, Chao Tang, and Kurt Wiesenfeld introduced the concept of *self-organized criticality* as a possible underlying mechanism leading to fractal characteristics in systems made up of large numbers of interacting parts, whether sand grains, molecules, or galaxies. According to its proponents, the theory provides a fundamentally different way of viewing a wide range of phenomena, including certain types of noise in electronic devices, turbulence during fluid flow, the pattern of energy released during earthquakes, the ebb and flow of sunspot activity, the irregular flickering of radiation from distant quasars, the large-scale structure of the universe, and the erratic fluctuations of stock-market prices and other economic indicators.

Bak and his colleagues contend that a complex dynamical system can naturally evolve into what they call a self-organized critical state, which is far from equilibrium and barely stable. Such a precariously balanced system always operates on the edge of collapse, yet resiliently responds to external stresses by returning in a characteristic way to a "critical" state.

One way to picture such a system is to imagine building a sandpile on a table by slowly adding grains of sand. At first, the pile is quite flat. Gradually, the sandpile gets steeper and steeper. Now and then, avalanches occur, and as the pile grows, the avalanches become bigger. Eventually, the sandpile reaches a critical state—a point at which the pile's slope no longer changes even though the pile itself gets bigger.

Anyone who has played with sand has seen this characteristic behavior. A bucket of dry sand dumped onto a table produces a cone-shaped mound. Whether the pile is small or large, its slope is always the same. Moreover, adding extra sand to an established pile triggers avalanches. By the time the avalanches die out, the pile's characteristic profile is back. In other words, the system regulates itself. Randomly adding sand grains to a pile triggers avalanches that bring the slope back to a particular angle known as the *angle of repose*.

Bak's theory predicts that when a sandpile is at its angle of repose, adding more sand grains or slightly tilting the pile's base would generate avalanches of all sizes, limited only by the pile's extent. Furthermore, the avalanches would occur at widely varying intervals. An observer sitting at some spot on a sandpile and measuring what is going on as a function of time and space would find features on all time scales and all length scales.

To test this idea, Bak and his colleagues constructed a simple computer model designed to capture some of the crucial features of sandpile behavior. In their theorist's view of a sandpile, each square of a two-dimensional grid has a certain numerical value representing the local slope (see Figure 4.17). When a "sand grain" lands on a square, the square's value goes up by 1. If that value happens to exceed a defined critical value, then sand starts to flow, and the value goes down by 1. That shift affects the

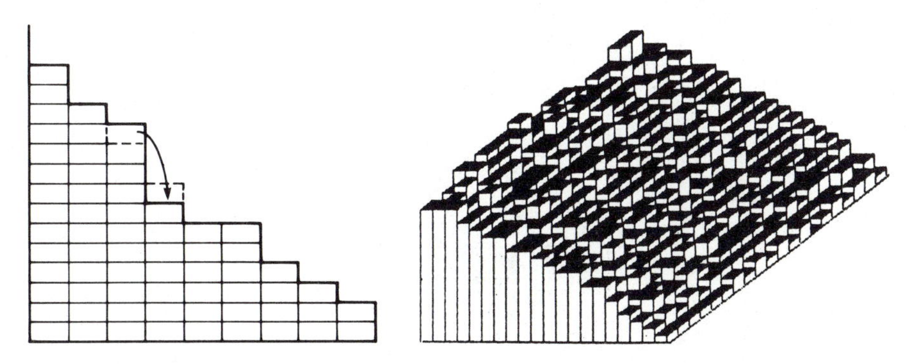

Figure 4.17 In a simple, one-dimensional model of a sandpile, grains topple when the difference in height between two neighboring columns exceeds a specified value. That shift sometimes forces other grains to slide (*left*). This model can be generalized to two and higher dimensions (*right*).

square's nearest neighbor, which may then also exceed the critical value, and so on. The result is a chain reaction that continues until no further shifts are possible.

Although the model is incredibly simple, it captures much of the complexity of critical phenomena already known in physics. Computer simulations show that adding single grains to a sandpile in its critical state produces avalanches of many different sizes. Sometimes the arrival of a sand grain has no effect. At other times, the arrival of one unit of sand causes a slide, and that in turn forces two neighbors to slide, producing a three-unit avalanche. Occasionally, very large avalanches occur (see Figure 4.18). It's also possible to monitor how much sand is flowing at a given time. That flow is very irregular.

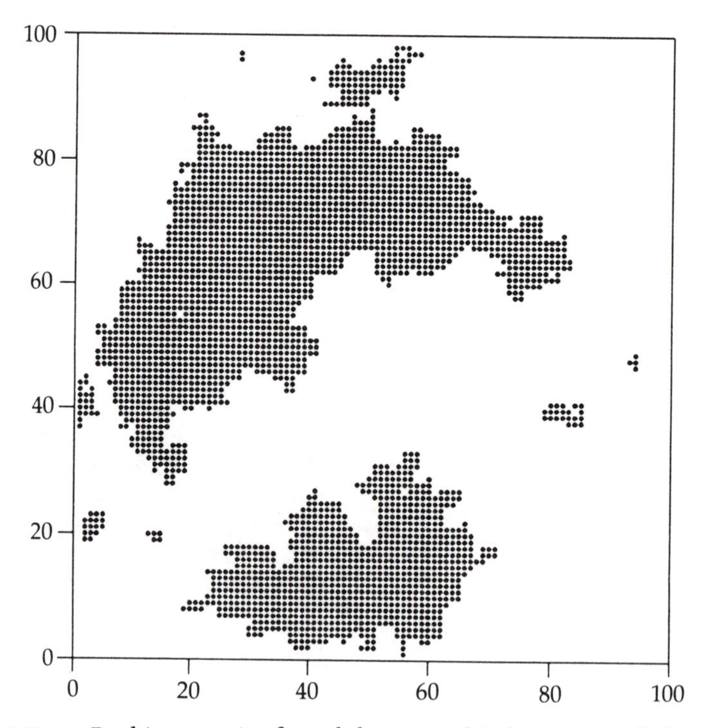

Figure 4.18 Pushing a unit of sand downward induces a sand slide. Such a shift often triggers further slides, shown as variously sized dark clusters on this two-dimensional array representing a sandpile seen from above.

Merely by changing the language, the same computer model seems to work just as well for earthquakes. In this case, the computer model represents colliding tectonic plates that make up the earth's crust, and the critical slope corresponds to a critical pressure. Thus, an earthquake is a chain reaction triggered by a local instability that propagates like a collapsing row of dominos through a geological fault system (see Figure 4.19).

The model also predicts that small changes in the system can have widespread effects. Its behavior is extremely sensitive to details far away from the starting point. If this, indeed, represents the physics of earthquakes, then earthquakes are unpredictable in the deepest possible way.

Bak's analogy between sandpile avalanches and self-organized criticality prompted several researchers to take a closer look at the behavior of real sandpiles. Sand is such a commonplace material that many people think they have a good idea of what happens in a sandpile. Yet the behavior of sand is complex and poorly understood. Dry sand can be heaped into a pile that retains its shape, like a solid. But if the pile is disturbed or its slope becomes too steep, sand grains flow downhill. A closer look,

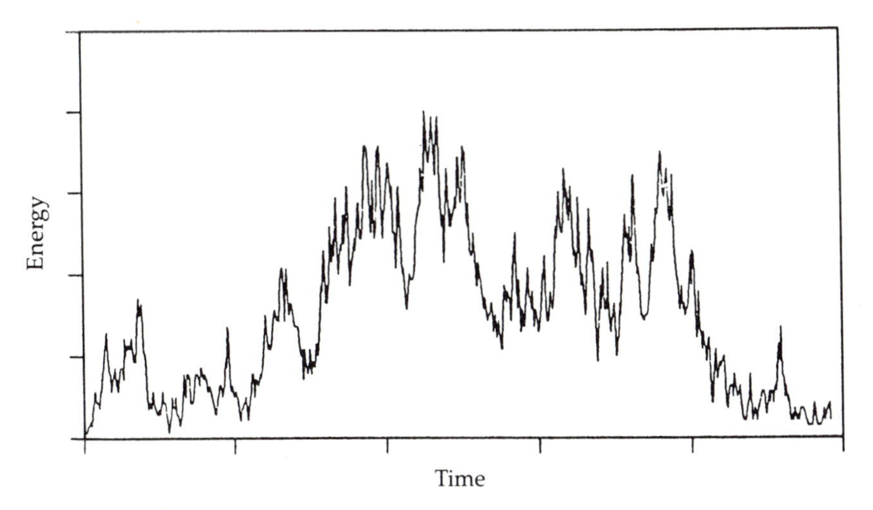

Figure 4.19 A computer simulation, based on a sandpile model, mimics the pattern of energy released during an earthquake.

however, reveals the flow to be more like that of a dense gas of heavy, colliding particles than a true liquid. Such multifaceted behavior compounds the difficulties of investigating and understanding sand behavior.

To study sandpile behavior, one group of researchers, led by physicist Sidney Nagel, observed avalanches in a slowly rotating, cylindrical drum half filled with sand. They discovered that nothing happens until the sandpile slope reaches an angle a few degrees higher than the angle of repose. Then the whole slope suddenly shifts. The avalanches are almost always enormous. Moreover, depending on the angle and rotation speed of the apparatus, the avalanches happen at particular times.

In the traditional picture of sandpile avalanches, the angle of repose represents a balance between friction and gravitational force. Avalanches start when the gravitational force on an exposed sand grain is larger than the amount of friction between sand grains. The new results show that for a sand grain on a surface to start moving, it must be able to roll or slide over its neighbors just below. Therefore, sand starts moving when the pile's slope is higher than the angle of repose.

In other words, sandpiles behave in ways that are not as simple or as obvious as many scientists had assumed. A sandpile doesn't collapse until its slope reaches a threshold angle, at which point the whole slope collapses, producing a global landslide that brings it back to the angle of repose. The simple, intuitive picture of the angle of repose as the critical angle of the system has to be modified, and the direct analogy between the dynamics of sandpiles and physical systems exhibiting a critical point is not well founded.

However, such negative experimental results haven't slowed the theoretical exploration of self-organized criticality. Although the sandpile analogy may be flawed, other physical phenomena may still show this effect.

Bak argues that his concept of self-organized criticality provides a possible explanation for "$1/f$" noise, where f stands for frequency. This simple formula expresses how the intensity of this type of sound at any given frequency depends upon the value of the frequency. Such noise has been detected in electronic devices and fits the kind of irregular fluctuations seen in the luminosity of certain stars and in river flows measured over long periods of time. This type of noise, although irregular, is not random. For example, if the light from a quasar is measured over a long period, you see features that last for a very long time and other features that last for a very short time. Taken together, such oscillations have no typical time scale (see Figure 4.20).

Bak and other researchers have developed a number of computer models to test the applicability of this concept to a variety of

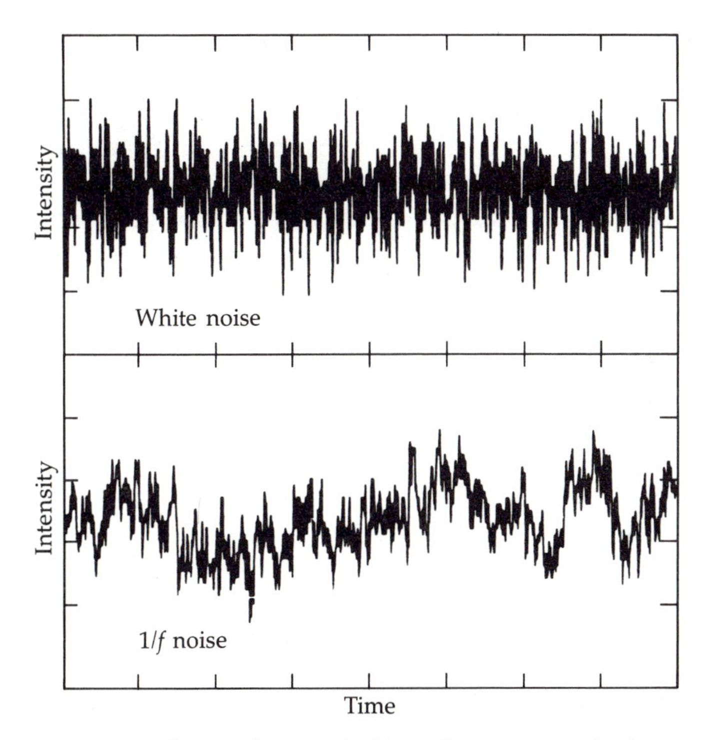

Figure 4.20 Noise, the random variations of some quantity in time, has a number of different forms. For white noise, the most random, the frequency, f, doesn't depend at all on intensity (*top*). On the other hand, $1/f$ noise, a commonly found type of fluctuation in nature, does depend in a specific way on its intensity.

complex, interactive dynamical systems. Because realistic models of turbulence or of other motions in meteorology and geology would be too complicated to make, researchers create simple models that capture only a few features of real physical systems. The idea is to look for mathematical quantities—indices or exponents in mathematical expressions describing characteristics of the model—that are universal: the same numbers appear consistently for a wide variety of models. Such universal quantities may reflect fundamental properties that apply not only to simple model systems but also to realistic situations.

One of Bak's models resembles the way a fire might spread through a forest if a madman randomly set trees ablaze. A computer works with a grid in which each square is marked with a 0 (no tree), 1 (a tree), or 2 (a burning tree). After each turn, burning trees die, and neighbors of burning trees catch fire on the subsequent turn, and then die. Fires propagate, but fresh trees appear randomly in unoccupied squares (see Figure 4.21).

Starting with a random pattern of fires scattered over the entire grid, the system evolves to a critical state in which fires develop a stringlike structure. That leaves trees in clusters of many different sizes (see Color Plate 11).

Such a construct serves as a possible model for turbulence. Just as tree growth adds fuel to keep fires burning in characteristic patterns, energy uniformly pumped into a fluid by stirring or heating turns a uniform flow into a swirling pattern of self-similar vortices that have no typical size.

Bak has found that the game of Life, invented by mathematician John Conway, also evolves to a critical state. As in the fire model, the game takes place on a two-dimensional square lattice. Each square assumes a value of either 0, which indicates the absence of a live individual at that spot, or 1, which signifies the presence of a live individual. Whenever a live individual has four or more neighbors in surrounding squares, it dies on the next turn (from overcrowding). If a live individual has no neighbors or just one, it dies of loneliness. If a square, whether occupied or not, has exactly two nearest neighbors, nothing changes. An unoccupied square with three neighbors sees a birth.

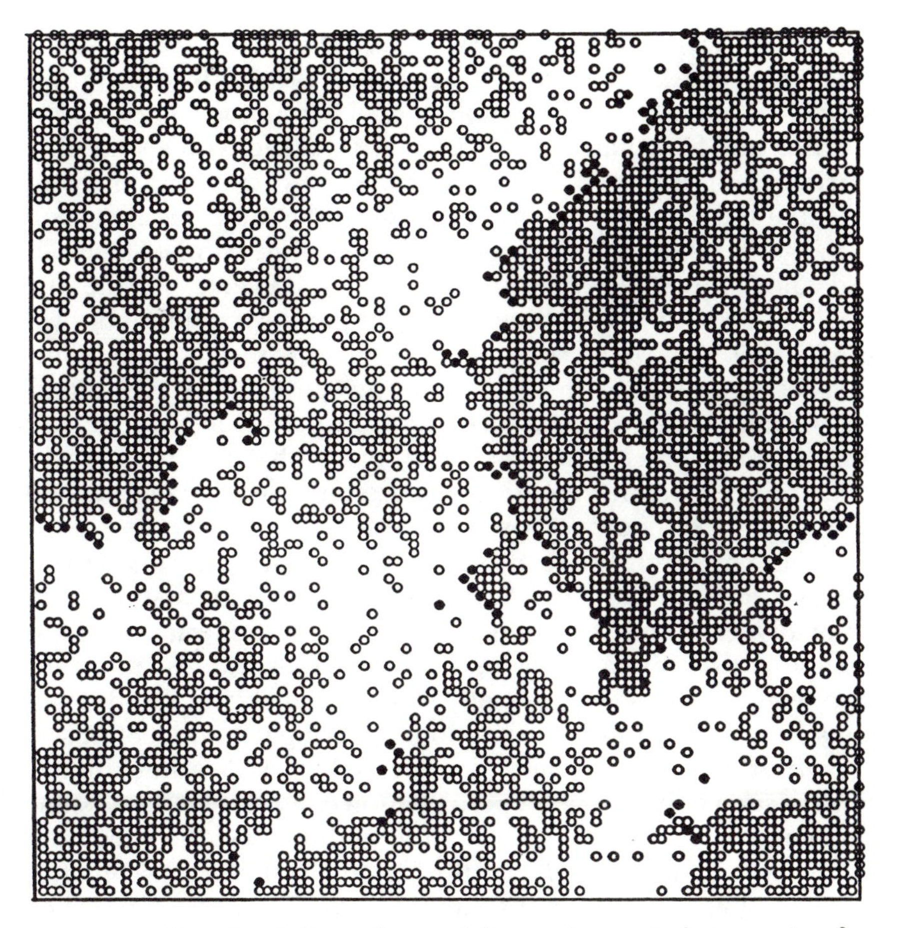

Figure 4.21 In a simple forest-fire model, trees (*open circles*) grow in a few empty sites at each time step. Trees on fire (*filled circles*) burn until the next time step, spreading flames to trees at their nearest-neighbor sites.

It doesn't matter very much what the precise rules are. The idea is to mark the squares randomly, then let the system evolve according to the rules until it comes to a stop. Live individuals are added to the grid in the same way that a few sand grains can be randomly added to a sandpile at its angle of repose. By monitoring the system's activity, researchers find that the game of Life operates at a critical point. They see a fractal structure in the pattern of clusters of live individuals. To Bak and others, the game is a grossly simplified but reasonable model for biological evolution.

Bak and his collaborators have also worked with a simplified version of the game of Life, constructed in order to identify the essential mechanisms in play and to see if the same features apply to the game on lattices in three and higher dimensions. The modified rules allow the researchers to look at "objects" on the grid that move about and collide, exploding and spreading out. The simulations show in a crude way how matters may condense into galaxies, then scatter in collisions between galaxies (see Figure 4.22). The model suggests that the universe itself may be in a critical state, which would account for its observed clumpiness, evident in the arrangement of galaxies into clusters and clusters into superclusters on very large scales.

Some researchers are also studying the possibility of applying the concept of self-organized criticality to economic systems. Traditionally, economists tend to describe systems in terms of simple equations that relate a few quantities such as interest rates and employment levels. They usually study the effects of small deviations from an equilibrium situation. Applying the concept of self-organized criticality to an economic system adds a new dimension. In such a state, a small perturbation can create either a small effect or a large one. There's no limit on how long the effect may last or how far it could extend through the system. These fluctuations are much stronger than those possible in an equilibrium model.

Simple computer models of economic systems, in which grid sites represent decisions made by individuals and simple rules define how information about those decisions spreads, in fact do evolve to a critical state. Such models of economic systems also demonstrate that self-organized critical systems are intrinsically unpredictable. In order to predict the fluctuations of economic indices, one must have complete information about the total system, which is impossible to achieve. Even the tiniest disturbance can sometimes have an enormous impact, which spreads through the whole system. Indeed, plotting the Dow-Jones stock-market

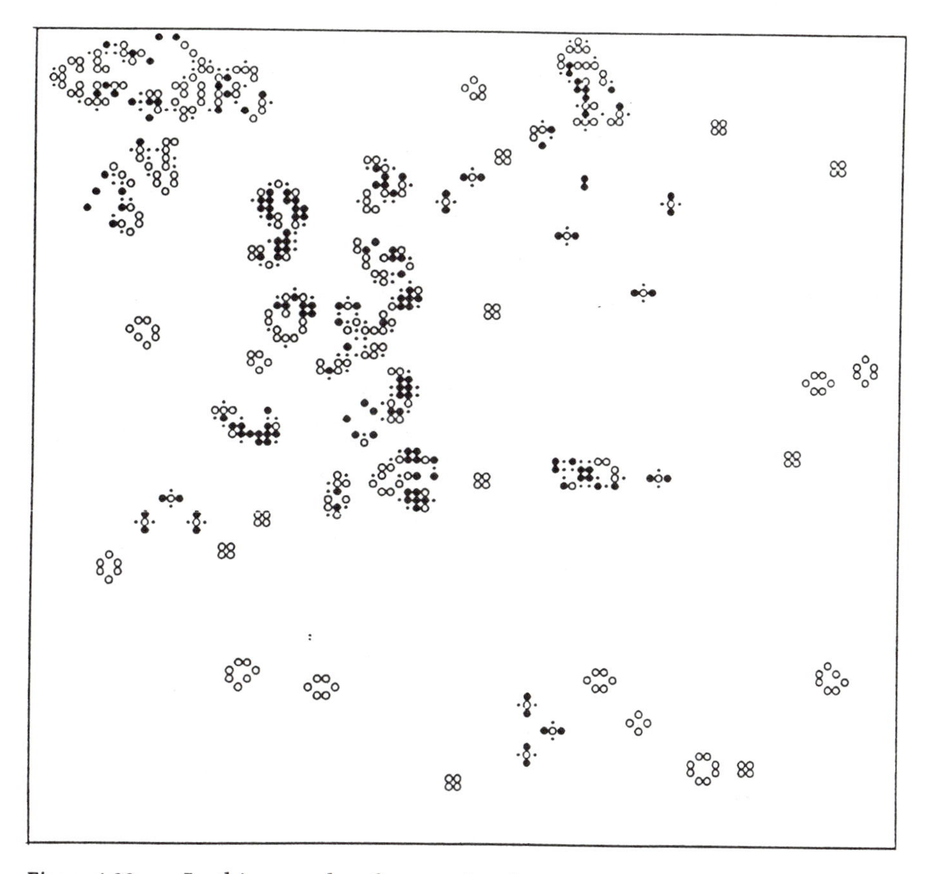

Figure 4.22 In this snapshot from a simulation involving the game of Life, open circles indicate live passive sites; filled circles represent dying active sites. The dots indicate sites where birth will take place at the next time step. The game's simple rules generate numerous configurations, representing the variety and complexity of local stable societies.

index reveals just such a fluctuating pattern having a wide range of temporal and spatial scales.

Two central questions about the theory of self-organized criticality remain unresolved. It isn't clear yet if a wide range of models showing this kind of behavior all really share certain quantifiable, "universal" characteristics. One group of scientists has shown that small changes in the rules governing computer models of avalanches can produce behaviors that don't all have the same general properties.

It also isn't clear if the concept of self-organized criticality is at all useful for describing and explaining the behavior of real systems. For example, the theory doesn't seem to apply to $1/f$ noises in electronic devices. Such devices don't show the fractal structures that Bak's theory suggests ought to give rise to $1/f$ noise.

Nevertheless, the concept has caught the attention of numerous researchers in many fields. Self-organized criticality may turn out to be yet another instance of a theoretical construct or a piece of abstract mathematics that sits around for a while before somebody stumbles across a situation to which it actually applies.

5

Number Play

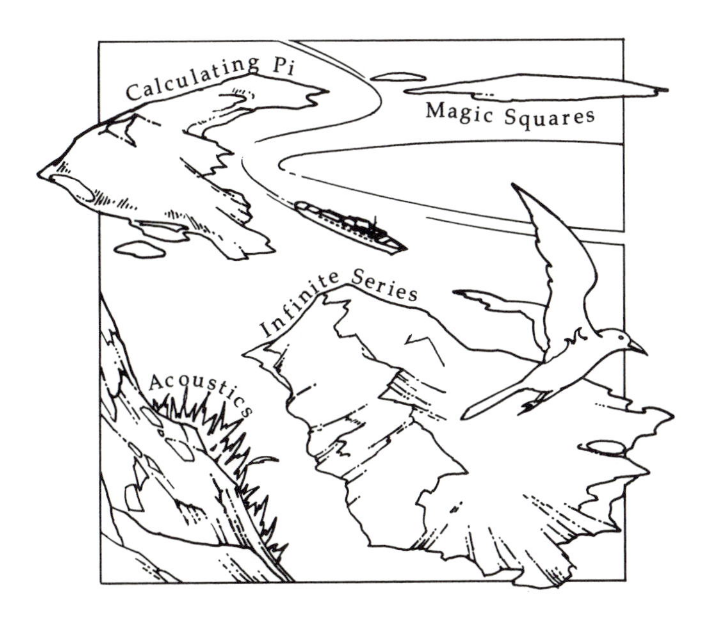

Have you heard the one about an itinerant entertainer traveling with a wolf, a goat, and a basket of cabbages? The showman comes to a river and finds a small boat that holds only himself and one passenger. For obvious reasons, he can't leave the wolf alone with the goat, or the goat with the cabbages. How does he get his cargo safely to the other side?

This well-known brainteaser has been around for centuries. In the thirteenth century, it featured three beautiful brides and their young, handsome husbands. All are jealous: the husbands of the wives and the wives of the husbands. Using a two-person boat, how many trips does it take to ferry them all across a river —without igniting a fit of illicit passion?

These two puzzles and their many variants are ways of dressing up a relatively straightforward mathematical problem. Since the days of ancient Egypt and Babylon, people have often used such devices to turn routine mathematical exercises into problems that tickle and challenge the mind.

Mathematical puzzles and games remain remarkably popular. Puzzle addicts throughout the world snap up many of the hundreds of such books published every year. A wide range of magazines feature puzzle columns. The appearance of a new, ingenious puzzle can stir up frenzied activity. In just three years, sales of Rubik's cubes grew beyond 100 million.

Amusement is one of humankind's strongest motivating forces. Although mathematicians sometimes belittle a colleague's work by calling it "recreational" mathematics, much serious mathematics has come out of recreational problems, which test mathematical logic and reveal mathematical truths. Philosopher and mathematician Bertrand Russell noted, "A logical theory may be tested by its capacity for dealing with puzzles, and it is a wholesome plan, in thinking about logic, to stock the mind with as many puzzles as possible, since these serve much the same purpose as is served by experiments in physical science."

Probability theory originated in questions about gambling. Ingenious solutions to tiling problems, as discussed in Chapter 3, are seen in the folk art of many cultures and in physical theory for describing crystals. Similar connections between recreational problems and important mathematical questions occur in number theory, geometry, graph theory, and many other areas of mathematics.

Mathematical physicist John L. Synge summed it up: "In submitting to your consideration the idea that the human mind is at its best when playing, I am myself playing, and that makes me feel that what I am saying may have in it an element of truth."

A Shortage of Small Numbers

Canadian mathematician Richard K. Guy is a collector. He patiently and painstakingly searches far and wide for the unexpected and quirky among the family of whole numbers. He looks for unusual patterns.

Identifying patterns and asking the right questions are two of the most important ingredients of mathematical research. Lamentably, there's no foolproof recipe for generating good questions and no formula for recognizing whether an observed pattern will lead to a significant new theorem or is merely a lucky coincidence. Until a mathematical proof is constructed to settle the question, a mathematician must rely on fallible, empirical evidence.

Consider the remarkable sequence of integers 31, 331, 3331, 33331, 333331, 3333331. Each of these integers is a prime number, that is, divisible only by itself and the number one. Is the sequence's next number, 33333331, a prime? The answer is yes. Unfortunately, the pattern falls apart with the succeeding number, 333333331, which turns out to be the product of 17 and 19,607,843. A promising pattern is slain by a cruel counterexample.

Guy's specimens are all instances of sequences that depend on the whole-number value, *n*, of some parameter. In the first example, n represents the number of threes in each integer. The pattern works for $n = 1, 2, 3, 4, 5, 6$, and 7, but fails when $n = 8$. For any sequence that depends on the value of n, experience shows that sometimes a pattern persists, but frustratingly often the pattern is simply a figment of the smallness of the values of n for which the example has been worked out.

For many years, Guy has been trying to encapsulate his findings in the form of a universal law. So far, the best he can manage is the statement: "There aren't enough small numbers to meet the many demands made of them." He calls it the Strong Law of Small Numbers.

"It is the enemy of mathematical discovery," Guy argues. "When you notice a mathematical pattern, how do you know it's for real? We are easily led astray by spurious patterns, which do not continue as the numbers get larger. On the other hand, genuine patterns are often hidden by a few exceptions near the beginning."

As an instance of the misleading behavior of small numbers, Guy cites the fact that 10 percent of the first 100 numbers are perfect squares (1, 4, 9, 16, 25, 36, 49, 64, 81, and 100). On the basis of this pattern, one could conjecture that 10 percent of the first 1,000 numbers would also be perfect squares, but some quick calculations show the conjecture to be ill-founded. Only about 3 percent are perfect squares.

On the other hand, the statement that all prime numbers are odd is almost true. The only exception occurs right at the beginning. In a sense, as Guy points out, two is the "oddest" prime.

Guy's tussles with such aberrant numerical behavior have led him to formulate an important, elegantly simple theorem: "You can't tell by looking." The theorem, he insists, "has wide application, outside mathematics as well as within," and it can be "proved by intimidation."

Many of Guy's examples, gathered from numerous sources, concern prime numbers. One of the most famous involves numbers of the form $P = 2^{2^n} + 1$. When $n = 0$, $P = 2^{2^0} + 1 = 2^1 + 1 = 2 + 1 = 3$, a prime number. For $n = 1$, $P = 2^{2^1} + 1 = 5$, another prime. For $n = 2$, $P = 17$; for $n = 3$, $P = 257$; for $n = 4$, $P =$

65,537. The numbers 3, 5, 17, 257, and 65,537 are all primes. Does the pattern continue? Mathematician Pierre de Fermat thought so when, more than three centuries ago, he proposed that all numbers of the form $2^{2^n} + 1$ are prime. Alas, when $n = 5$, the number is not a prime but the product of 641 and 6,700,417. The Strong Law strikes again.

A tastier problem concerns slicing a round cake into pieces —not in the conventional way but in a fashion that probably only a mathematician would find appetizing (see Figure 5.1). The idea is to define a certain number of points, n, along the cake's rim, then to slice the cake so that the cuts join all possible pairs of points. The question is how many separate pieces of cake are created by the cuts.

The answer for $n = 1$ (one point on the rim) is, of course, one. With only one point, no cuts can be made. When $n = 2$, a cut joins two points, dividing the cake into two pieces. For $n = 3$, the number of pieces, p, is four; for $n = 4$, $p = 8$; for $n = 5$, $p = 16$. The sequence 1, 2, 4, 8, 16 looks familiar. Does the pattern hold for larger numbers of points? The answer is no. The number of pieces for $n = 6$ is 31. The sequence continues: 57, 99, 163, 256, 386, 562, 794, 1,093, . . . However, it is possible to work out a formula that gives you every term in this sequence: $(n^4 - 6n^3 + 23n^2 - 18n + 24)/24$.

Pennies show up in a low-budget hexagon construction project (see Figure 5.2). Seven pennies can be laid out to form a hexagon in which each side is two pennies long. A hexagon with each side made up of three pennies consists of a total of 19 pennies. As the length of the hexagon's side goes from 1 penny to 5 pennies, the total number of pennies involved in each case is 1, 7, 19, 37, and 61. The members of this sequence are called "hex" numbers. Interestingly, $1 + 7 = 8$, $8 + 19 = 27$, $27 + 37 = 64$, $64 + 61 = 125$. Each of these partial sums appears to be a perfect cube. For example, $8 = 2^3 = 2 \times 2 \times 2$, $27 = 3 \times 3 \times 3$, and so on. Does this pattern continue when larger hexagons built from pennies are included?

The pattern is genuine. It's handy to regard the nth hex number as comprising the three faces at one corner of a cubic stack of

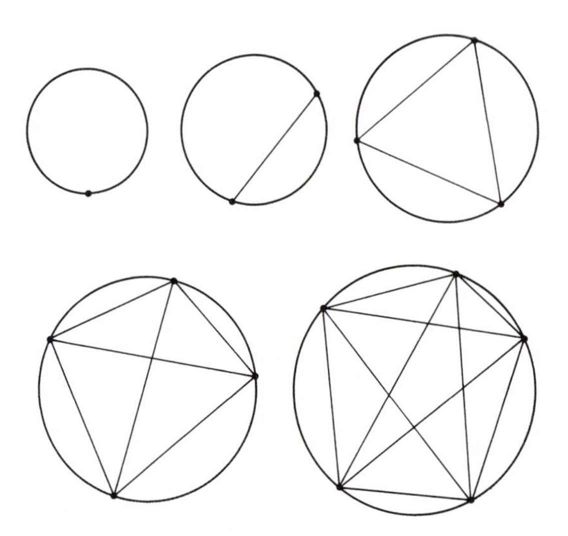

Figure 5.1 Joining in every possible way a number of points marked on a circle slices the circle into a certain number of regions. How does the number of regions depend on the number of points?

n^3 unit cubes. Expressed algebraically, the $(n + 1)$th hex number, $1 + 6 + 12 + \ldots + 6n$ (or equivalently, $3n^2 + 3n + 1$), when added to n^3, gives $(n + 1)^3$.

Another problem involves writing down a string of positive integers, say, from 1 to 11 (although you can go as high as you like). Cross out every second number; that is, all the even numbers. Then add up the remaining numbers, writing down the answers, or partial sums, along the way. The resulting sequence (1, 4, 9, 16, 25, 36) consists of consecutive squares of positive integers. Does the pattern continue if the string has more than 11 positive integers? Yes. This is a mathematically proper way of generating a sequence of squares.

What happens if you delete every third number, compute the partial sums, then delete every second partial sum, and calculate new partial sums? This time (see Figure 5.3), the resulting sequence consists of consecutive cubes.

Guy's collection of problems illustrates the major role that disinformation in the form of misleading patterns plays in the

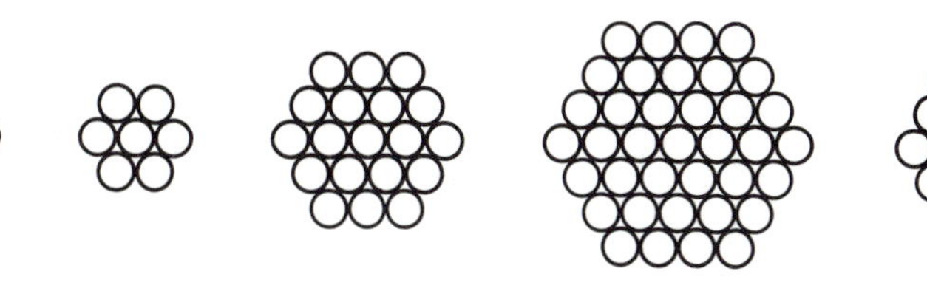

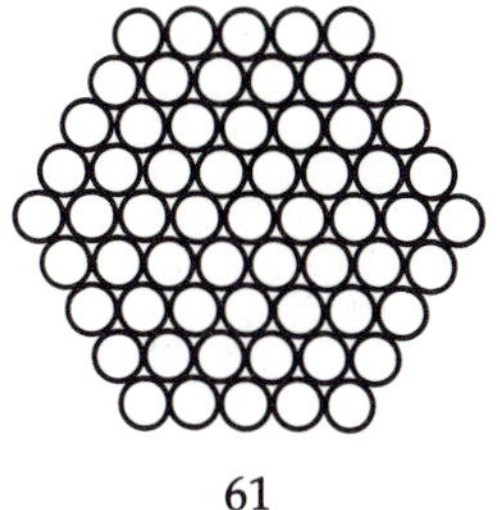

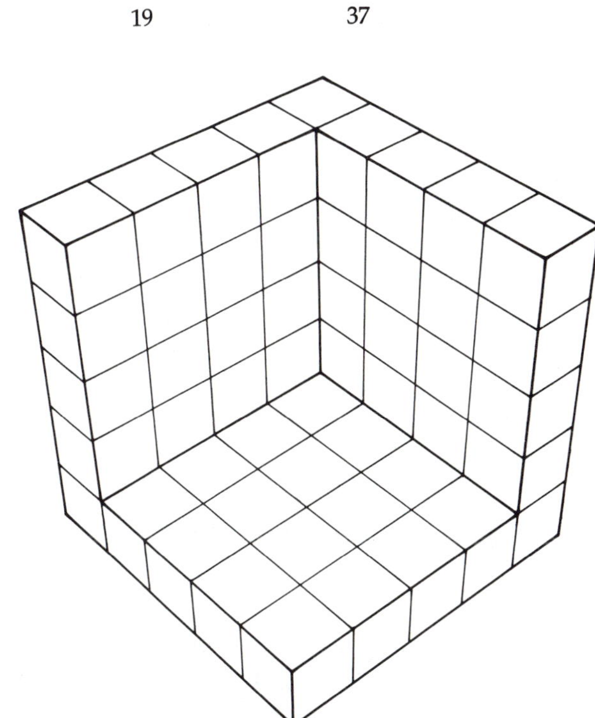

Figure 5.2 Adding together the so-called hex numbers (*top*) produces the partial sums 1, 8, 27, 64, 125, which appear to be perfect cubes. To explain this behavior, it's useful to regard the nth hex number as comprising the three faces at one corner of a cubic stack of n^3 unit cubes (*bottom*).

pursuit of mathematical truth. He and his fellow collectors could fill many volumes with examples of how the Strong Law of Small Numbers has led to significant mathematical theorems, or has misled investigators into looking for theorems that don't exist. It's

1 ~~2~~ 3 ~~4~~ 5 ~~6~~ 7 ~~8~~ 9 ~~10~~ 11 ...

1 4 9 16 25 36

1 2 ~~3~~ 4 5 ~~6~~ 7 8 ~~9~~ 10 11 ~~12~~ 13 14 ~~15~~ 16 ...

1 ~~3~~ 7 ~~12~~ 19 ~~27~~ 37 ~~48~~ 61 ~~75~~ 91

1 8 27 64 125 216

Figure 5.3 Write down the positive integers, delete every second number, and form the partial sums of those remaining. The result is a sequence of perfect squares (*top*). This time, delete every third number, calculate the partial sums, then delete every second partial sum. The result is a sequence of perfect cubes (*bottom*).

all part of the trial-and-error (often mainly error) effort that characterizes much of mathematical research.

Tickling the Mind

Recreational mathematics furnishes an extensive playing field for both amateur and professional mathematicians. It offers both a sense of play and the joy of discovery. Sometimes the results are mathematically trivial; surprisingly often, they lead to new mathematical insights that, like the sphere-packing problem in Chapter 3, turn out to be useful. Moreover, it's startling how often simply stated problems turn out to be deceptively difficult, if not impossible, to solve.

Curious properties sometimes lurk hidden within seemingly undistinguished numbers. Take the story concerning Indian mathematician Srinivasa Ramanujan: when a mathematician and friend visiting Ramanujan remarked that the taxi by which he had arrived had a "dull" number—1729, or $7 \times 13 \times 19$—

Ramanujan replied that 1,729 is actually an extraordinary number. It's the smallest number expressible as a sum of two cubes in two different ways. Ramanujan noted that both $1^3 + 12^3$ and $9^3 + 10^3$ equal 1,729, which proved his point and showed his marvelous facility with numbers and their properties.

Several years ago, mathematician Albert Wilansky, when calling his brother-in-law, noticed that the telephone number had a striking property. The number, 493-7775 (4,937,775), is a composite number, and can be expressed as the product of prime numbers: $3 \times 5 \times 5 \times 65{,}837$. Interestingly, when the digits of the original number are added together, the result (42) equals the sum of the digits in the prime factors ($3 + 5 + 5 + 6 + 5 + 8 + 3 + 7 = 42$). This discovery marked the birth of "Smith" numbers, named for Wilansky's brother-in-law.

The smallest Smith number is 4 because the number's factors, 2×2, when added together also equal 4. The next one is 22; then comes 27. Overall, there are 376 Smith numbers among the first 10,000 positive integers. About 3,300 Smith numbers lie between zero and 100,000, and a slightly smaller number are between 100,000 and 200,000. The largest known Smith number is more than 2.5 million digits long.

Investigations by both amateur and professional mathematicians have revealed that special patterns of digits automatically produce Smith numbers, but no one has found a method of generating every possible Smith number. In 1985, mathematician Wayne McDaniel managed to prove that there is an infinite number of Smith numbers. But many other questions about Smith numbers remain unanswered.

Heiko Harborth is one of a number of mathematicians who play with matches. "Matchsticks are among the cheapest and simplest objects for puzzles," he says. "Whole books have been devoted to matchstick puzzles." One group of matchstick problems involves constructing patterns in which a given number of sticks meet end to end, without crossing each other, at every point in a geometric figure on a flat surface. For example, a figure made up of three sticks laid out as an equal-sided triangle has 2 sticks meeting at each corner. Three sticks is the smallest number that can be used to create a pattern in which 2 sticks meet at every vertex (see Figure 5.4, left). The problem is tougher when 3 sticks must meet at every corner. The answer requires a figure made up of a minimum of 12 sticks that meet, 3 at a time, at 8 vertices (see Figure 5.4, right).

No one has found the answer for 4 sticks meeting at each vertex. The smallest construction found so far is Harborth's arrangement of 104 matchsticks meeting at 52 points (see Figure 5.5). No one knows whether it's the smallest construction that meets the criteria. It is known, however, that no analogous patterns exist for five or more sticks meeting at each vertex.

Playing with matchsticks raises questions that pertain to the mathematical field of *graph theory*—the study of ways in which points can be connected. Graphs often play important roles in circuit and network design, and some types of graphs can be explored with matchsticks. For example, a line segment is the only possible figure using a matchstick to connect two points. With two or three matchsticks, there are two configurations connecting three points: the points fall in a line or a triangle. There are 5 different arrangements that connect 4 points, 13 arrangements for connecting 5 points, and 50 for connecting 6 points (see Figure 5.6). But no one has found a general formula for determin-

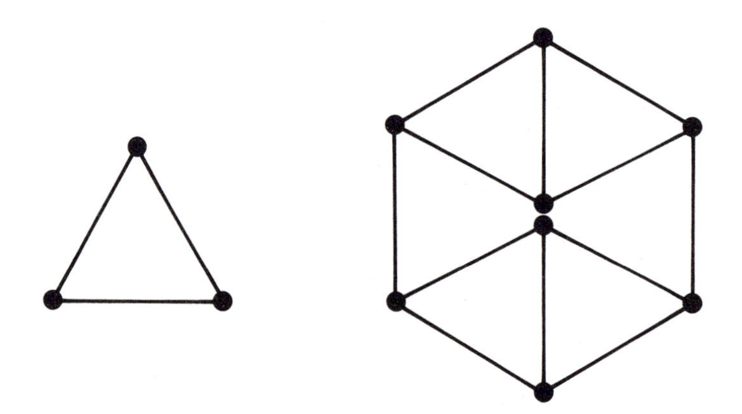

Figure 5.4 These are the solutions to the problem of arranging the smallest possible number of matchsticks in the plane so that two matchsticks meet at every vertex (*left*) and three matchsticks meet at every vertex (*right*).

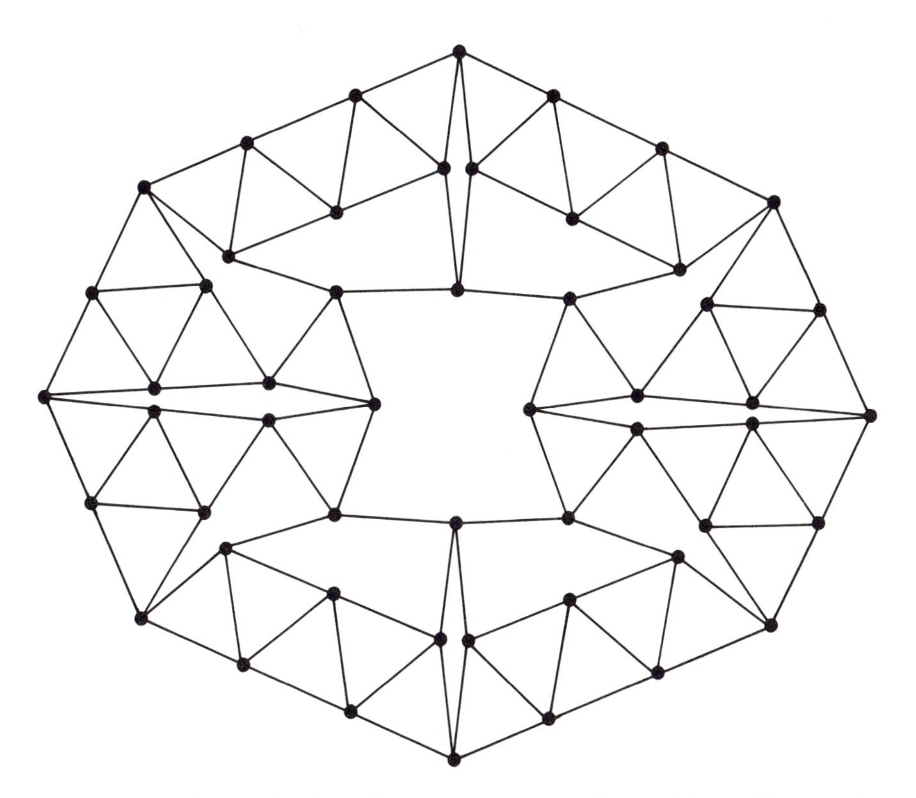

Figure 5.5 This is the best known answer to the problem of arranging matchsticks so that four sticks meet at each vertex. No one knows whether the problem can be solved with the use of fewer matchsticks.

ing the number of possible arrangements for any given number of points.

Magic squares have fascinated people for thousands of years. They consist of a set of whole numbers arranged in a square so that the sum of the numbers is the same in each row, in each column, and along each diagonal. Some magic squares have special properties, such as using only consecutive numbers. In ancient China, a three-by-three square that uses all of the digits from 1 to 9 was said to bring good luck.

Puzzle enthusiast Lee Sallows has developed a new category of magic-square problems. He started with an inscription written

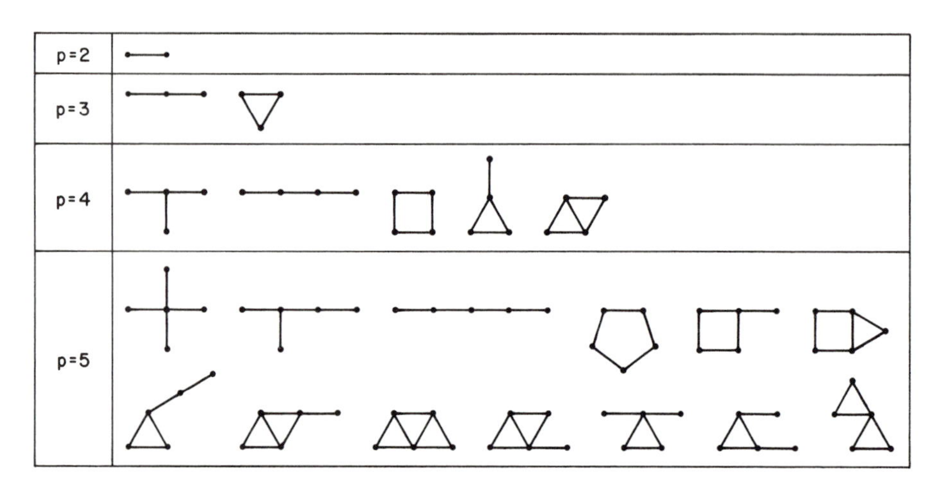

Figure 5.6 There is only one way in which a matchstick can join two points, two ways for matchsticks to connect three points, and five ways to connect four points. As the number of points increases, the number of possible matchstick patterns escalates rapidly.

in the runic alphabet (see Figure 5.7). A translation of the inscription reveals a three-by-three magic square with an unusual property. When the number of runic characters that make up each number in the original magic square is written down in each space, the derived numbers also form a magic square! Moreover, this second magic square is made up of consecutive integers from 4 to 11. Curiously, the same pattern also works in modern English.

Sallows' discovery led to a search for other examples, which in many different languages, have the same property. For columns, rows, and diagonals totaling less than 200, French has only one such magic square, whereas English has more than seven. Welsh, on the other hand, has more than 26. For totals less than 100, none occurs in Danish, but six occur in Dutch, 13 in Finnish, and an incredible 221 in German. There's even a three-by-three English square from which a magic square can be derived, which in turn yields a third magic square.

The search has now been expanded to four-by-four and five-by-five language-dependent magic squares. Sallows describes his quest as "a search for ever more potent magic spells."

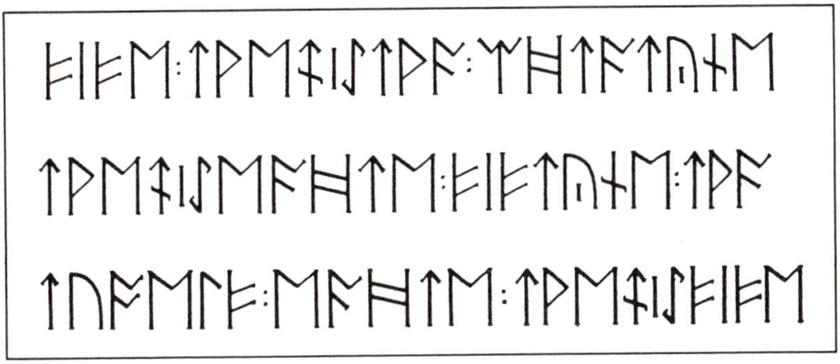

5	22	18
28	15	2
12	8	25

→

five	twenty-two	eighteen
twenty-eight	fifteen	two
twelve	eight	twenty-five

→

4	9	8
11	7	3
6	5	10

Figure 5.7 The deciphering of a runic inscription (*top*) led to the discovery of a new class of magic squares. The inscription consists of a set of numbers that fill the spaces in a three-by-three square. The numbers in each column, row, and diagonal add up to the same number, namely 45 (*bottom left*). Remarkably, when the number in each space is replaced by the number of letters in the word for the number, a new magic square is created—one that uses all the numbers from 3 to 11 (*bottom right*). It works in both the original language and in modern English (*bottom middle*).

Anyone who has gone through the chore of tiling a floor or installing a suspended ceiling can appreciate how much easier the job is if the tiles or panels evenly fit the area to be covered. For example, panels that are 4 feet long and 2 feet wide nicely cover a ceiling that happens to be, say, 6 feet by 8 feet. All the space can be filled without cutting any panels. In algebraic terms, if an $a \times b$ rectangle is tiled with copies of a $c \times d$ rectangle, then c divides evenly into either a or b, and d divides the other. The theorem, in a form that applies in higher dimensions as well, was proved about 20 years ago by Dutch mathematician N. G. de Bruijn.

Mathematicians subsequently suggested and proved a more general theorem: Whenever a larger rectangle is tiled by smaller rectangles, each of which has at least one integer side, then the

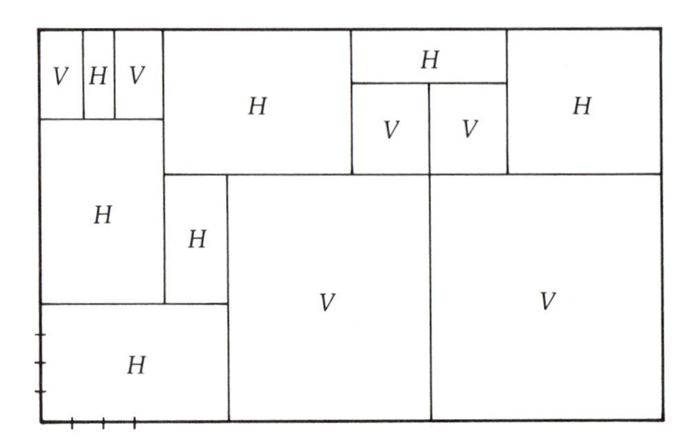

Figure 5.8 An example of a tiling in which each tile has at least one integer side. The tiles labeled *H* have integer width; those labeled *V* have integer height.

tiled rectangle has at least one integer side (see Figure 5.8). The original proof for this theorem, as in the case of de Bruijn's theorem, required the use of complicated mathematics involving double integrals and complex numbers. It was like using a cannon to kill a mouse.

In 1985, mathematician Hugh Montgomery discussed the theorem at a meeting and stimulated a search for a more elementary proof. That search subsequently turned up 13 alternative proofs, bringing a wide range of different mathematical techniques to bear on the problem.

Richard Rochberg came up with one of the simplest proofs. Each rectangular tile is divided into a checkerboard pattern comprised of squares in which each square is half a unit wide. Because each tile has an integer side, it carries an equal amount of black and white. If such tiles completely cover a large rectangle, then the rectangle must also have equal amounts of black and white. Therefore, at least one of the sides of the large rectangle has an integer length. If the rectangular tiles don't completely cover the large rectangle, the large rectangle can be split into four pieces, three of which have equal amounts of black and white while the fourth does not (see Figure 5.9).

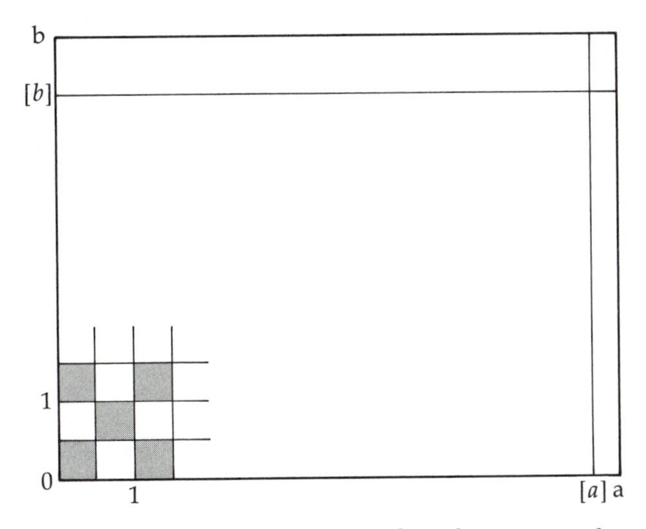

Figure 5.9 If neither *a* nor *b* is an integer, then the upper right corner has more black than white.

Other proofs involve double integrals with real instead of complex numbers, various ways of counting squares, and the use of prime numbers, polynomials, step functions, or graph theory. Although several of the proofs are mathematically related and most have similar ingredients, important differences show up when the methods are tried on variations of the original theorem. What if the tiles are like flexible postage stamps pasted on the surface of a cylinder or on a doughnut-shaped form? What happens in higher dimensions, when n-dimensional bricks are stacked in n-dimensional boxes? Some methods of proof are more powerful than others because they yield more general results. But deciding which proof is the "best" isn't easy because that depends on the criteria used to define "best." Quite possibly, the "best" proof hasn't yet been found.

Mathematician Solomon Golomb introduced the term *polyomino* to describe shapes that cover connected squares on a checkerboard (see Figure 5.10). There is only one type of monomino and domino, but there are 2 trominoes, 5 tetrominoes, and 12

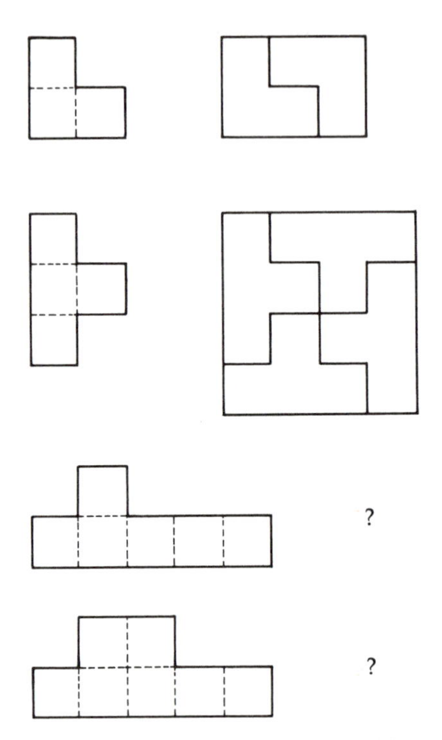

Figure 5.10 One type of polyomino tiling problem involves showing whether any number of tiles in the shape of a given polyomino can be fitted together to form a rectangle. Such rectangular arrangements have been found for many polyominoes, including trominoes and tetrominoes (*top*). Finding such an arrangement for the hexomino and heptomino shown (*bottom*) proved to be difficult.

pentominoes. As the number of connected squares increases, the number of different possible configurations grows rapidly.

Polyominoes are now the basis for thousands of mathematical puzzles. One of the more intriguing of such puzzles involves proving that polyominoes of a certain shape can be laid down to form a complete rectangle. Mathematicians have proved that the general problem of whether a given polyomino can tile a rectangle can't be decided because no computer program designed to test whether polyominoes of an arbitrary shape could be fitted together to form a rectangle could provide a yes or no answer within a reasonable time. There are just too many possible arrangements to check (see Chapter 6).

Even small polyominoes pose problems. With square and rectangular pieces, the problem is easy to solve. But a piece made up of six squares, connected so that five lie in one row with a sixth attached to the row's second square, is much more difficult to deal with. So is the case of a polyomino consisting of seven squares, with five in one row, and two in the second and third places in an adjacent row. Until recently, despite a great deal of effort, all anyone knew was that these two polyominoes could be used to tile a "semi-infinite" strip and that copies of each would fill a finite rectangle, except for a small square hole somewhere inside (see Figure 5.11).

In 1987, software engineer Karl Dahlke, combining perseverance with clever computer programming, managed to solve both problems. Dahlke, who is blind, heard the problem in the audio edition of a magazine. After trying out various possibilities, he decided to program his personal computer to search for an answer systematically. Dahlke's computer is equipped with a speech synthesizer that converts the computer's output into sound. Only after adding several programming tricks designed to circumvent time-consuming situations, in which the computer was trapped in endlessly repeating patterns, could Dahlke find his solutions (see Figure 5.12). It turns out that the size of the solutions is a bit beyond what one could easily do by hand, but well within the range of a personal computer.

Even in a subject as elementary as plane euclidean geometry, which has roots going back thousands of years, an astonishing amount remains unknown. Dentist and amateur mathematician Leon Bankoff has spent many years exploring the "intuition-shattering" geometric properties of a figure called the shoemaker's knife (see Figure 5.13, top). This geometric figure, first described by Archimedes and known to the ancient Greeks as the *arbelos*, has for centuries been a rich source of curious and unexpected discoveries.

The figure consists of three semicircles that touch each other in pairs, all having their diameters on a common line. Over the

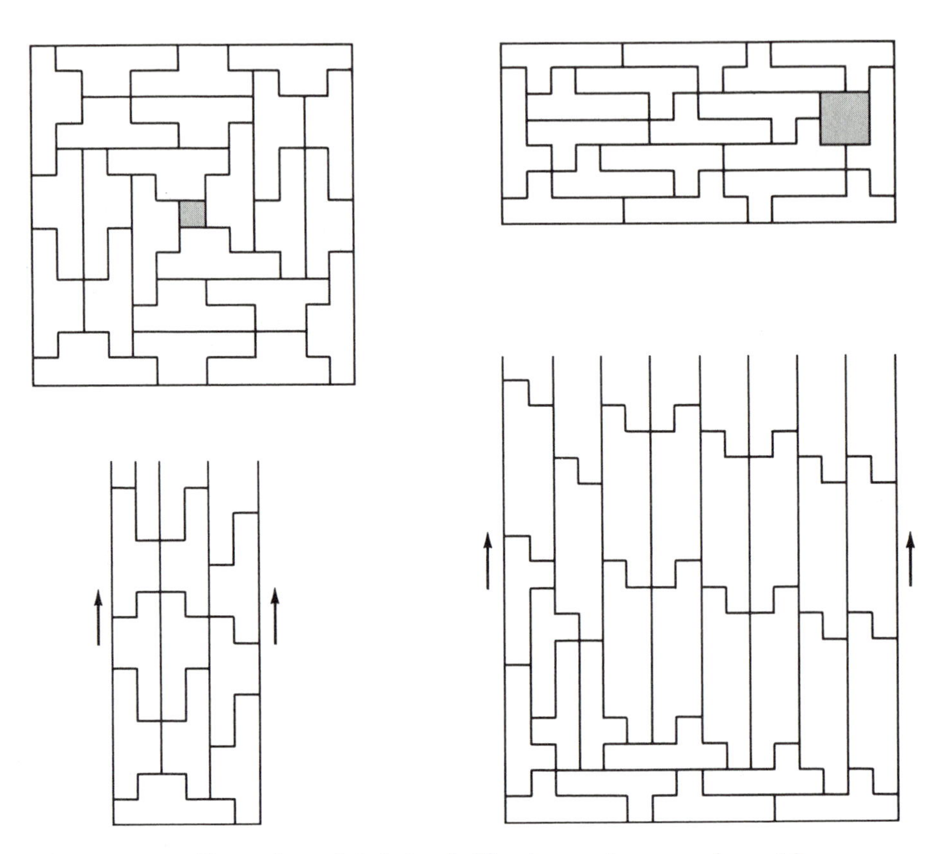

Figure 5.11 Examples of infinite half-strips and rectangles with square holes made from hexominoes and heptominoes.

years, its geometric properties have attracted the attention of mathematical luminaries no less than Pappus, Descartes, Fermat, and Newton. Yet, despite all this attention, certain features on the arbelos remained unnoticed until Bankoff himself found them. In particular, Archimedes had proved that the figure contains a pair of inscribed circles (see Figure 5.13, bottom). These circles came to be called the twin circles of Archimedes. Bankoff managed to show that the figure contains a third, identical inscribed circle. Archimedes' twins are actually two members of a set of triplets.

Figure 5.12 Dahlke's solutions to the problem of forming a rectangle with either hexominoes or heptominoes.

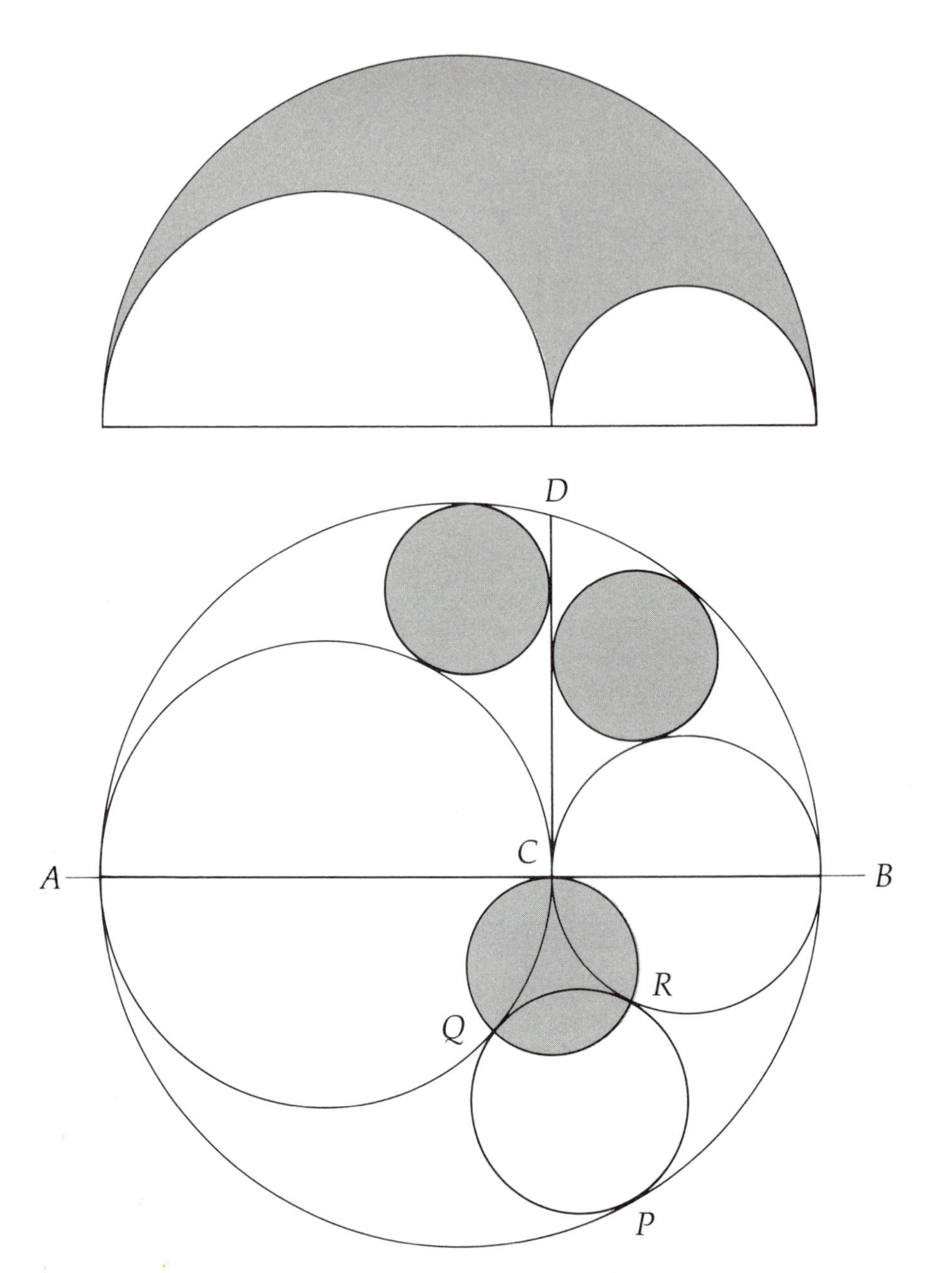

Figure 5.13 The shoemaker's knife, or arbelos (*shaded area, top diagram*), first described centuries ago by Archimedes, has a host of striking geometric properties, many of which have gone unnoticed for years. Amateur mathematician Leon Banoff discovered a third circle, *CQR*, drawn for clarity within a mirror image of the shoemaker's knife, that is equal in size to the long-known twin circles shown on either side of the line *DC* (*bottom*).

Not all mathematical research needs to dip into arcane notation and mind-numbing mental gymnastics. Recreational mathematics offers a wide range of problems, many of which have fascinated untold generations of both amateur and professional mathematicians. The tools are generally simple—often just pencil and paper. What's required is a lot of care, patience, and persistence, the same qualities that make for good mathematical research.

The Formula Man

The work of Indian mathematician Srinivasa Ramanujan (see Figure 5.14) represents mathematical play at its highest, most serious level. Born in 1887 in the small town of Erode and raised in Kumbakonam in southern India, Ramanujan lived most of his life in obscurity and poverty. Largely self-educated and hooked on mathematics, he filled three notebooks, between 1903 and 1914, with page after page of mathematical formulas and relationships. He came up with perhaps as many as 4,000 theorems, all stated without rigorous proof.

Although Ramanujan died in 1920 at the age of 32, his work is still the subject of considerable interest. Recent results such as the computation of π to millions of digits and the use of certain mathematical models in two-dimensional statistical mechanics rely completely or in part on Ramanujan's pioneering research.

Ramanujan's work reveals a genius for finding numerical patterns, hidden laws, and relationships in the wilderness of numbers. No one really knows what led him to his astonishing array of mathematical discoveries or how he proved his results. He spent most of his life far from centers of mathematical activity and worked in his own way, drawing formulas and theorems from a mental landscape unconnected with the frontiers of contemporary mathematics. Paper was too precious for him to use to write down anything but final results. He experimented and tested his

Figure 5.14 Srinivasa Ramanujan, born in 1887 in India, managed despite a limited formal education to derive numerous original formulas and theorems.

formulas on a slate, erasing all his intermediate work, recording only the final result on paper.

Ramanujan's jottings are important enough that several mathematicians have spent the last decade or so going through his notebooks, systematically proving each of his theorems. His notebooks, although intended strictly for his own use, contain very few serious errors.

Ramanujan's life as a professional mathematician began in 1914, when, at the age of 17, he accepted an invitation from the prominent mathematician G. H. Hardy to come to Cambridge University in England. The year before, Ramanujan had sent sam-

9 Wintry Forest

10 Snow Carpets

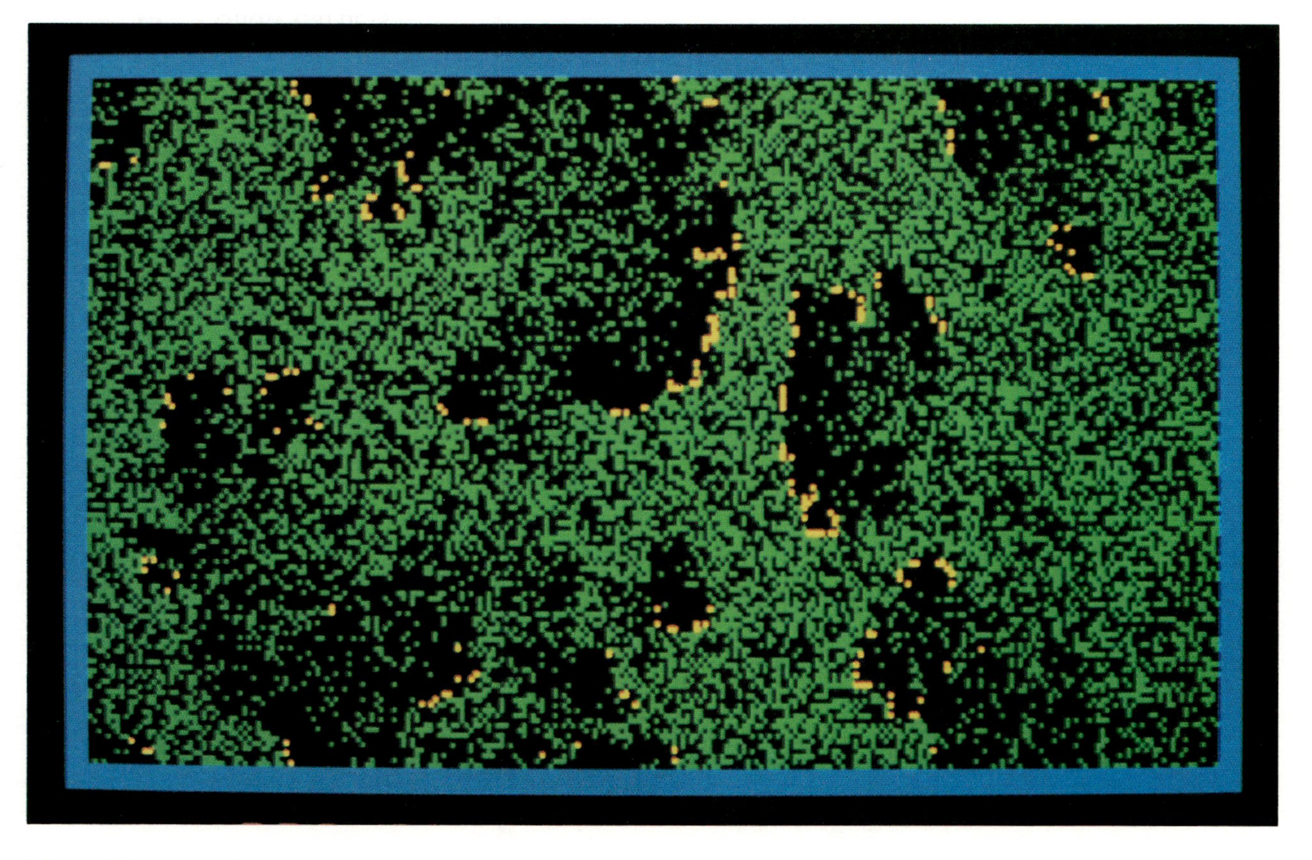

11 Playing with Fire

12 Pi-Scape

13 Pendulum Spaghetti

14 A Chaotic Question

15 San Marco Wonderland

16 Four-Dimensional Addition

ples of his work to several mathematicians in England, including Hardy, who had been impressed with the formulas and theorems. "I had never seen anything like them before," Hardy wrote in an essay years later. "A single look at them is enough to show that they could only be written down by a mathematician of the highest class. They must be true because, it they were not true, no one would have had the imagination to invent them."

Ramanujan spent 5 years in England, publishing a considerable number of papers and achieving worldwide fame. In 1917, he contracted a mysterious, incurable disease. He spent the last year of his life back in India. Speaking about that final year, Ramanujan's widow has said: "He was only skin and bones. He often complained of severe pain. In spite of it, he was always busy doing his mathematics. That, evidently helped him to forget his pain. I used to gather the sheets of paper which he filled up. I would also give the slate whenever he asked for it."

Unlike much of mathematics, which is carefully built up piece by piece on foundations laid by the work of earlier mathematicians, Ramanujan's contributions are unique. After more than 70 years, a large proportion of Ramanujan's accomplishments has not been rediscovered or duplicated. Even the 600 or so results from the last year of Ramanujan's life contain intriguing assertions that are not yet proved.

Much recent effort has gone into editing Ramanujan's notebooks (see Figure 5.15) and the 130 pages of scribbled material (now known as the "lost" notebook) from the last year of his life. In a kind of mathematical archaeology, mathematicians are carefully examining all of Ramanujan's formulas and theorems one by one. If a formula or theorem seems new to mathematics, they try to prove it. If a formula or theorem is already known, they look for its original source.

More than a decade ago, mathematician George Andrews discovered stored away at Cambridge University's Trinity College library the pages of Ramanujan's nearly illegible, unlabeled jottings from his fateful final year. In the intervening years, Andrews has managed to decipher and prove many of the manuscript's theorems. "[Ramanujan] was able to pick out and touch upon functions of great interest," says Andrews. "As we study them, we

326.

If β be of the 3rd degree,

i. $\sqrt[8]{\frac{\alpha^3}{\beta}} - \sqrt[8]{\frac{(1-\alpha)^3}{1-\beta}} = \sqrt[8]{\frac{(1-\beta)^3}{1-\alpha}} - \sqrt[8]{\frac{\beta^3}{\alpha}} = 1.$

ii $\sqrt[4]{\alpha\beta} + \sqrt[4]{(1-\alpha)(1-\beta)} = 1.$

iii. $m = 1 + 2\sqrt[8]{\frac{\beta^3}{\alpha}}$ and $\frac{3}{m} = 1 + 2\sqrt[8]{\frac{(1-\alpha)^3}{1-\beta}}$

iv. $m^2\left(\sqrt[8]{\frac{\alpha^3}{\beta}} - \alpha\right) = \sqrt[8]{\frac{\alpha^3}{\beta}} -$ [illegible]

v. $m = \frac{1 - 2\sqrt[8]{\frac{\beta^3(1-\beta)^3}{\alpha(1-\alpha)}}}{1 - 2\sqrt[4]{\alpha\beta}} = \sqrt{1 + 4\sqrt[8]{\frac{\beta^3(1-\beta)^3}{\alpha(1-\alpha)}}}$ and

$\frac{3}{m} = \frac{2\sqrt[8]{\frac{\alpha^3(1-\alpha)^3}{\beta(1-\beta)}} - 1}{1 - 2\sqrt[4]{\alpha\beta}} = \sqrt{1 + 4\sqrt[8]{\frac{\alpha^3(1-\alpha)^3}{\beta(1-\beta)}}}$

vi. If $\alpha = p\left(\frac{2+p}{1+2p}\right)^3$ then $\beta = p^3 \cdot \frac{2+p}{1+2p}$ so that

$1 - \alpha = (1+p)\left(\frac{1-p}{1+2p}\right)^3$ & $1 - \beta = (1+p)^3 \cdot \frac{1-p}{1+2p}$

vii. $m^2 = \sqrt{\frac{\beta}{\alpha}} + \sqrt{\frac{1-\beta}{1-\alpha}} - \sqrt{\frac{\beta(1-\beta)}{\alpha(1-\alpha)}}$ and hence

$9/m^2 = \sqrt{\frac{\alpha}{\beta}} + \sqrt{\frac{1-\alpha}{1-\beta}} - \sqrt{\frac{\alpha(1-\alpha)}{\beta(1-\beta)}}$

viii. $\sqrt[8]{\alpha\beta^5} + \sqrt[8]{(1-\alpha)(1-\beta)^5} = 1 - \sqrt[8]{\frac{\beta^3(1-\alpha)^3}{\alpha(1-\beta)}}$

$= \sqrt[8]{\alpha^5\beta} + \sqrt[8]{(1-\alpha)^5(1-\beta)} = \sqrt{\frac{1 + \sqrt{\alpha\beta} + \sqrt{(1-\alpha)(1-\beta)}}{2}}$

ix. $\sqrt{\alpha(1-\beta)} + \sqrt{\beta(1-\alpha)} = 2\sqrt[8]{\alpha\beta(1-\alpha)(1-\beta)}.$

$m^2\sqrt{\alpha(1-\alpha)} + \sqrt{\beta(1-\beta)} = 9/m^2 \cdot \sqrt{\beta(1-\beta)} + \sqrt{\alpha(1-\alpha)}.$

x. $m\sqrt{1-\alpha} + \sqrt{1-\beta} = \frac{3}{m}\sqrt{1-\beta} - \sqrt{1-\alpha} = 2\sqrt[8]{(1-\alpha)(1-\beta)}$ and

$m\sqrt{\alpha} - \sqrt{\beta} = \frac{3}{m}\sqrt{\beta} + \sqrt{\alpha} = 2\sqrt[8]{\alpha\beta}.$

xi. $m - \frac{3}{m} = 2\left\{\sqrt[4]{\alpha\beta} - \sqrt[4]{(1-\alpha)(1-\beta)}\right\}$ and

$m + \frac{3}{m} = 4\sqrt{\frac{1 + \sqrt{\alpha\beta} + \sqrt{(1-\alpha)(1-\beta)}}{2}}$

xii If $P = \sqrt[8]{16\alpha\beta(1-\alpha)(1-\beta)}$ and $Q = \sqrt[4]{\frac{\beta(1-\beta)}{\alpha(1-\alpha)}}$, then

$Q + \frac{1}{Q} + 2\sqrt{2}\left(P - \frac{1}{P}\right) = 0$

Figure 5.15 Ramanujan's "notebooks" were personal records in which he jotted down many of his formulas, almost always without proof or even a hint of where the formulas came from. The page shown contains various third-order modular equations—all in Ramanujan's nonstandard notation.

realize how revolutionary and how intriguing they are, and the real implications they have for mathematics in the current day."

Mathematicians laboring to prove theorems Ramanujan knew to be true are rewarded not so much for the results themselves but rather for the new techniques they need to develop to prove a result or derive a formula. By reconstructing where Ramanujan's formulas came from, mathematicians catch glimpses of general principles new to mathematics. Whole subdisciplines within mathematics have blossomed from these efforts.

The topic that Ramanujan loved best is infinite series. Mathematically, a series is simply the sum of a specified number of terms. Usually, a simple algebraic formula indicates what each successive term should be. For example, the series $1 + 4 + 9 + 16$ can also be written as the sum of all terms n^2, where n runs through the integers from 1 to 4. In the case of an infinite series, the summation process extends to integers of an arbitrarily large size.

One of Ramanujan's infinite series is now the basis for methods used to compute pi (π) to millions of decimal places:

$$\frac{1}{\pi} = \frac{\sqrt{8}}{9801} \sum_{n=0}^{\infty} \frac{(4n)!}{(n!)^4} \frac{[1103 + 26390n]}{396^{4n}}$$

In this case, the Greek letter sigma signifies summation, and the numbers above and below the symbol indicate that the formula is exactly correct only if all terms, starting with $n = 0$ and going to infinity, are added together. An exclamation mark means that the designated integer is actually the product of all positive integers up to and including the integer shown (that is, $4! = 1 \times 2 \times 3 \times 4 = 24$).

Ramanujan's equation arrives at values of π to large numbers of decimal places more rapidly than just about any other known infinite series. Each extra term in the summation adds roughly eight digits to the decimal expansion of π. In one of his papers, Ramanujan gives 14 other series for $1/\pi$, and, as usual, offers little explanation of where they came from. Even now, with the help of a more profound theoretical understanding and aided by new mathematical tools such as computer software for manipulating algebraic expressions, mathematicians still find it hard to generate the kinds of identities that Ramanujan apparently found so readily.

Ramanujan also expended a great deal of effort inventing and exploring structures called continued fractions. In this area, Ramanujan's work is probably unsurpassed in all of mathematical history.

In a sense, *continued fractions*, as the name hints, are fractions of fractions of fractions of fractions. . . . The following expression for $\pi/4$ is a typical example of one of these typographical nightmares:

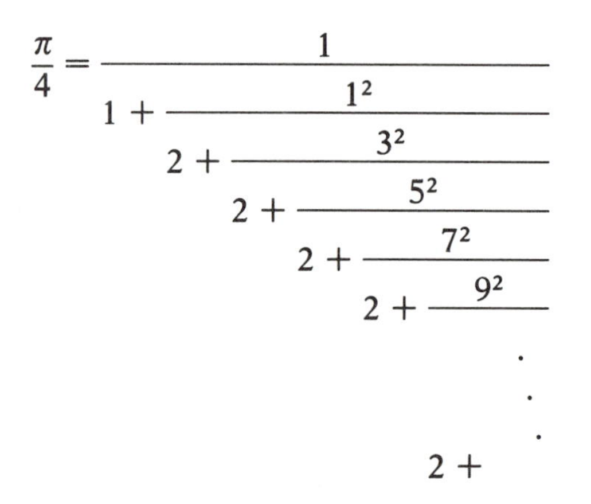

$$\frac{\pi}{4} = \cfrac{1}{1 + \cfrac{1^2}{2 + \cfrac{3^2}{2 + \cfrac{5^2}{2 + \cfrac{7^2}{2 + \cfrac{9^2}{\ddots\; 2 +}}}}}}$$

Each successive term in a continued fraction expansion sits deeper and deeper within the fraction's denominator.

Some of Ramanujan's best work is in the theory of partitions, which concerns how many ways a given whole number can be written as a sum of smaller whole numbers. For example, the integer 5 can be written out as a sum of smaller whole numbers in seven different ways (see Figure 5.16). Ramanujan noticed that whenever an integer leaves a remainder 4 after division by 5, then the number of its partitions is evenly divisible by 5. For example, the integer 9 has 30 distinct partitions, a number evenly divisible by 5. Similarly mysterious relationships hold for the divisors 7 and 11 but for no other integers. The proof for this mathematical behavior is now known to be extremely tricky, and before Ramanujan, nobody had suspected that partitions might have arithmetical properties of this kind.

In a way, Ramanujan's fascination with formulas was old-fashioned. The great period for the discovery of such relationships was in the eighteenth and nineteenth centuries, in the days of Leonhard Euler and Carl Friedrich Gauss. Now mathematicians are beginning to realize that Ramanujan, rather than being a hundred

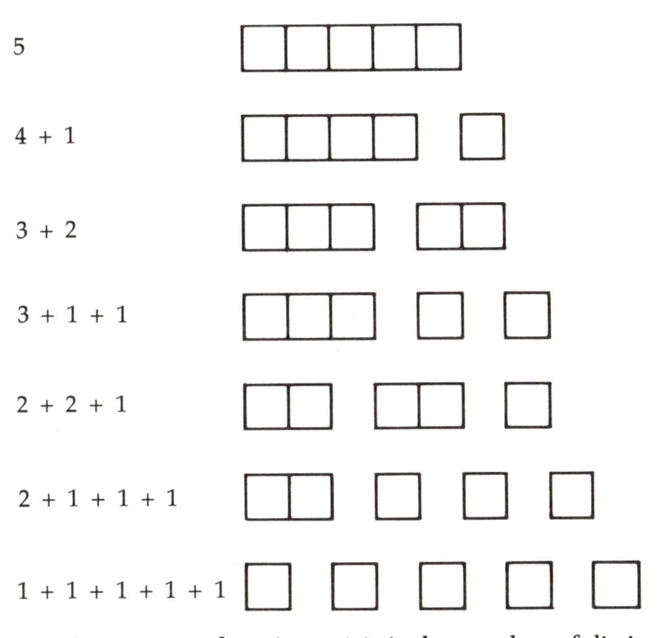

Figure 5.16 The partition function, $p(n)$, is the number of distinct ways in which n can be written as a sum of smaller or equal integers. For example, 5 can be expressed in seven different ways. Hence, $p(5) = 7$. The problem is equivalent to finding the number of different ways in which five blocks can be grouped in a row.

or so years late, was perhaps many decades ahead of his time. Some of his basic theories are just becoming understood and extended mathematically and, in some cases, applied to physical theory.

Some physicists have recently used Ramanujan's findings concerning partitions to solve problems in statistical mechanics. Using the so-called hard-hexagon model, in which interacting particles are laid out on a honeycomblike grid, physicists can study theoretically, say, the behavior of a layer of helium atoms sitting in a neat arrangement on a carbon surface. Some aspects of Ramanujan's work may also apply to string theory, which tries to provide a unified picture of matter and energy in the universe.

Most mathematicians believe they discover mathematics rather than invent it, so they think that in the long run, it doesn't really matter what an individual mathematician does. If one mathematician doesn't discover something, another will probably discover it within 50 years. The one possible counterexample in the twentieth century is Ramanujan. Mathematicians who have looked closely as his body of work find his unique contributions hard to fit into the rest of mathematics.

Had Ramanujan not died so young, his presence in modern mathematics might be even more immediately felt. Had he lived to have access to computers with special programs for manipulating algebraic quantities, who knows how much more spectacular his already astonishing career might have been. Even now, Ramanujan's legacy will keep mathematicians busy for many more decades.

In summing up Ramanujan's work, Hardy said, "He combined a power of generalisation, a feeling for form, and a capacity for rapid modification of his hypotheses, that were often really startling."

Pieces of Pi

The nature and meaning of the number pi (π) has intrigued and fascinated mathematicians for centuries. They know that π, the ratio of a circle's circumference to its diameter, is an *irrational number*. That means it can't be written out fully as a simple fraction. The fraction 22/7, for instance, is merely a crude approximation of π's actual value.

Moreover, π is also what mathematicians term a *transcendental number*. It can't be the solution, or root, of any algebraic equation with integer coefficients. In order words, π can never be evaluated exactly by any combination of simple operations involving the addition, subtraction, multiplication, or division of positive integers, or the extraction of square roots. Nevertheless, π can be expressed by formulas—for example, as the sum of an infinitely long series: $\pi = 4/1 - 4/3 + 4/5 - 4/7 + 4/9 - \ldots$ Using such a series, anyone, in principle, can calculate π to as many decimal places as desired (see Figure 5.17).

For most purposes, knowing the value of π to two decimal places is enough. In a few practical applications, someone may need to know the ratio to 10 or 15 decimal places, but rarely more. Thirty-nine decimal places suffice for computing the circumference of a circle girdling the known universe with an error no greater than the radius of a hydrogen atom.

Over the centuries, however, many amateur and professional mathematicians have taken up the challenge of calculating π to more and more decimal places. Much effort has gone into performing the tedious calculations required and looking for series that reach the desired answer faster. For some, it becomes an

π = 3.1415926535	8979323846	2643383279	5028841971	6939937510
5820974944	5923078164	0628620899	8628034825	3421170679
8214808651	3282306647	0938446095	5058223172	5359408128
4811174502	8410270193	8521105559	6446229489	5493038196
4428810975	6659334461	2847564823	3786783165	2712019091
4564856692	3460348610	4543266482	1339360726	0249141273
7245870066	0631558817	4881520920	9628292540	9171536436
7892590360	0113305305	4882046652	1384146951	9415116094
3305727036	5759591953	0921861173	8193261179	3105118548
0744623799	6274956735	1885752724	8912279381	8301194912
9833673362	4406566430	8602139494	6395224737	1907021798
6094370277	0539217176	2931767523	8467481846	7669405132
0005681271	4526356082	7785771342	7577896091	7363717872
1468440901	2249534301	4654958537	1050792279	6892589235
4201995611	2129021960	8640344181	5981362977	4771309960
5187072113	4999999837	2978049951	0597317328	1609631859
5024459455	3469083026	4252230825	3344685035	2619311881
7101000313	7838752886	5875332083	8142061717	7669147303
5982534904	2875546873	1159562863	8823537875	9375195778
1857780532	1712268066	1300192787	6611195909	2164201989 ...

Figure 5.17 The first 1,000 digits of the decimal expansion of pi (π).

obsession. Isaac Newton admitted as much in 1666 when he wrote, "I am ashamed to tell you to how many figures I carried these computations, having no other business at the time." He managed to compute at least 15 digits using the formula:

$$\pi = \frac{3\sqrt{3}}{4} + 24\left[\frac{1}{12} - \frac{1}{5 \cdot 2^5} - \frac{1}{28 \cdot 2^7} - \frac{1}{72 \cdot 2^9} - \cdots\right]$$

Nearly 2,000 years earlier, Archimedes had shown that π lies between 3 10/71 and 3 1/7. Using mathematical principles rather than direct measurements, he managed the feat by calculating the lengths of a pair of regular, or equal-sided, polygons, one fitting snugly inside a circle and the other outside (see Figure 5.18). His polygons had 96 sides. Sixteenth-century Dutch mathematician Ludolph van Ceulen dedicated much of his career to a computation based on the same method Archimedes used. He obtained a 32-digit estimate of π by calculating the perimeters of polygons having 2^{62} sides. Van Ceulen eventually reached 35 digits, and the last three—288—were engraved on his tombstone.

As mathematicians discovered more efficient formulas for evaluating π, they proceeded to calculate an increasing number of digits, reaching 205 digits by 1844. William Shanks, an obscure British mathematician and the headmaster of a boarding school, spent 20 years calculating the value of π, laboriously multiplying by hand numbers hundreds of digits long. Eventually, he reached 707 decimal places and published the result in 1853. But a century later, a new calculation showed that Shanks had made an error, mistakenly writing a 5 rather than a 4 in the 528th place and causing all subsequent digits to be wrong.

The coming of computers added glamor to the centuries-old hunt for π. In 1949, the pioneering computer ENIAC reached 2,037 digits. By 1973, the world record stood at a million decimal places. Then supercomputers came into play, and every year since 1985 has seen even larger jumps in the number of computed digits.

Computer scientists are now using the calculation of π to show off their newest, largest, fastest, most sophisticated computers, while at the same time testing the computer's circuits and

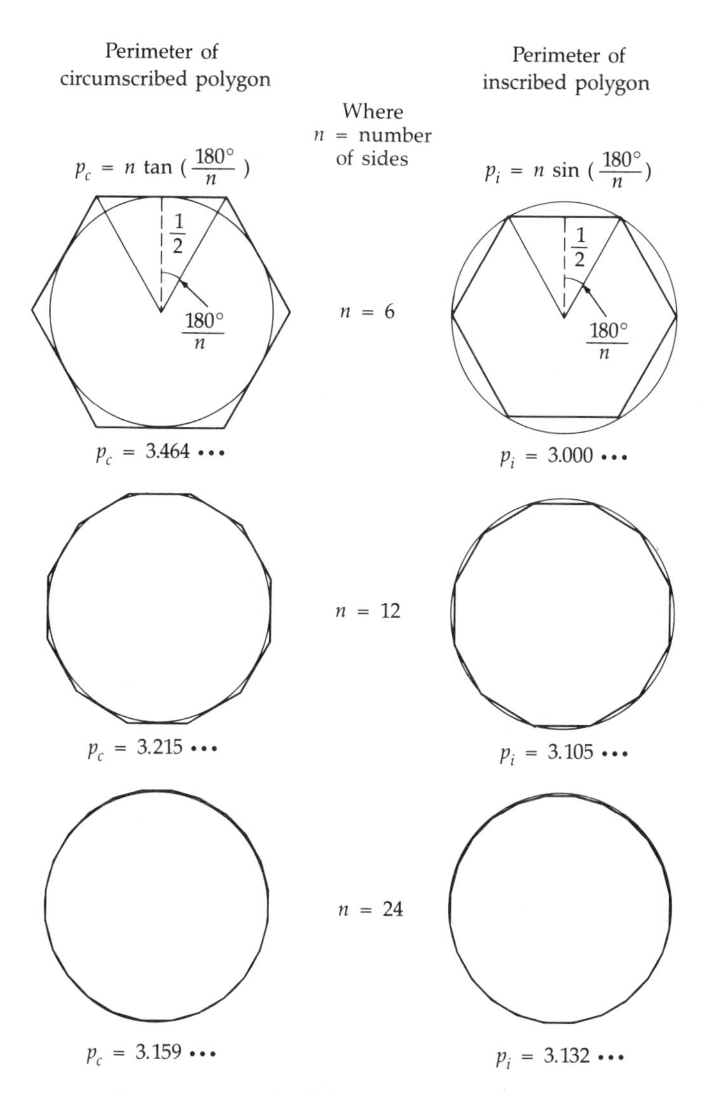

Figure 5.18 Archimedes' method for estimating π relied on inscribed and circumscribed polygons with sides of equal length on a circle having a diameter of one unit. The perimeters of the inscribed and circumscribed polygons served as lower and upper bounds for the value of π. The sine and tangent functions can be used to calculate the polygons' perimeters, as is shown here, but Archimedes had to develop equivalent relations based on geometric constructions. Using 96-sided polygons, he determined that π is greater than 3 10/71 and less than 3 1/7.

programs. Computing π gives such machines a thorough workout, pinpointing any subtle flaws. If even one error occurs in the computation, the result almost certainly would be completely in error after an initially correct section. An error in even one of the millions of digits of π would signal a problem in the computer or the computer program. On the other hand, if the result of the computation is correct, then the computer has done billions of operations without error. In effect, computing π is the ultimate stress test for a computer—a kind of digital cardiogram. Attempts to compute the digits of π also push computer scientists to devise new strategies for multiplying very large numbers rapidly and performing other numerical manipulations as efficiently as possible.

One reason for the mathematical interest in the decimal expression of π is that, in spite of its elementary nature, little is known about the mathematical properties of π. Furthermore, π turns up unexpectedly in myriad situations that have nothing to do with circles. It enters into numerous branches of mathematical analysis. For instance, the probability that a number picked at random from the set of integers will have no repeated prime divisors is 6 divided by the square of π. Moreover, a surprising number of infinite series has π in the answer. Indeed, it's merely a historical accident that we tend to define π geometrically as the ratio of a circle's circumference to its diameter. And researchers keep coming up with new formulas for π.

One issue of interest is whether all the digits from 0 to 9 occur equally often. A number with such an even distribution of digits is called "normal." Most mathematicians suspect that the sequence of decimal digits in π is normal, but this conjecture has never been proved. No one knows whether all digits from 0 to 9 appear infinitely often. There's always a chance that π may appear normal for a certain number of digits (now running into the billions) before hitting a pattern, such as $\pi = 3.1415926 \ldots 010010001000010000001 \ldots$. The question of π's normality only scratches the surface of the deeper question of whether the digits of π are random. This mystery is just one of many that encircle π.

Thus, there is continuing interest in performing statistical analyses on the decimal expansion of π to search for possible patterns (see Color Plate 12). A Japanese statistical analysis of the first 10 million digits shows no unusual deviation from expected behavior. On the average, the digits of π seem to be uniformly distributed (see Figure 5.19). But answering more general questions about the distribution of digits in the decimal expansion of π appears completely beyond the scope of current mathematical techniques. Mathematicians continue to compute digits of π partly because they have found very little else to do to understand the number.

So, the digit hunt continues. Its limiting factors are how much energy a programmer is willing to put into the hunt and how much time on a supercomputer a researcher can scrounge. If the pursuit were considered important enough by those who control computer time, billions of digits would be within reach.

Japanese computer scientist Yasumasa Kanada has shown the most persistence. In 1987, Kanada calculated π to 134 million digits. He reached 201,326,000 decimal places in 1988, and 536,870,000 digits in 1989. Each step up resulted from a combination of improvements in Kanada's basic π-computing program and the use of increasingly advanced supercomputers. To go further, Kanada is investigating ways of hastening the multiplication

Type of poker hand	Expected number	Actual number
No two digits the same	604,800	604,976
One pair	1,008,000	1,007,151
Two pairs	216,000	216,520
Three of a kind	144,000	144,375
Full house	18,000	17,891
Four of a kind	9,000	8,887
Five of a kind	200	200

Figure 5.19 Checking the distribution of the 2 million poker hands in the first 10 million decimal digits of π shows there is no significant deviation from the expected values.

of numbers up to a billion digits long. He's also looking for access to any of the new, experimental supercomputers being developed in Japan.

Kanada's motivation for pursuing π goes well beyond practical value. "It's like Mt. Everest," he says. "I do it because it's there." Because π is known to be an infinite decimal, there's no reason why Kanada can't continue his quest indefinitely, subject to the limits imposed by available computer technology and his own perseverance.

And there may be clever ways to speed up the computation. In 1985, Bill Gosper, one of the original computer hackers at MIT, followed an independent course, using a mathematical technique involving continued-fraction expansions and a small computer not especially built for numerical work. He managed to reach 17 million digits, still the largest continued-fraction expansion of π. Interestingly, Gosper used Ramanujan's continued-fraction formula for π at a time when no one had yet proved that Ramanujan's formula was correct. Ironically, Gosper's calculation helped to prove the correctness of Ramanujan's sum.

In 1989, two brothers, David and Gregory Chudnovsky, showed off a remarkable, alternative method for computing π. Their algorithm allowed them to reach 480 million decimal places and, a few months later, 1,011,196,691 digits in the decimal expansion of π. Printed out in line, the digits they computed would stretch nearly halfway across the United States.

The new formula, or identity, discovered and used by the Chudnovskys expresses π as a complicated sum. By evaluating more and more terms in such a sum, mathematicians get closer and closer to the true value of π. This particular identity converges to π faster than any other known formula.

But computing π quickly isn't enough. Computers don't work correctly all the time, and an accidental change even in one bit of data caused by a hardware fault, a speck of dust, or the destructive effects of a cosmic ray could completely corrupt a computation. So the Chudnovskys applied a "compute but verify" strategy. To avoid losing or corrupting painstakingly assembled data, the Chudnovsky method incorporates automatic verification and, where necessary, correction of faulty data. In effect, the verification scheme provides the equivalent of a complete set of fingerprints of a stranger you've never seen. It doesn't help you find the stranger, but you'll know him when you meet him.

Using computer programs written in the computer language FORTRAN, the Chudnovskys tested their method on two different computers, calculating π to 480 million decimal places on a Cray-2 and an IBM 3090. It was the first massive calculation of π done on two machines with totally different architectures. The computations were performed at odd times over a period of 6 months, sometimes in segments only 15 minutes long. In the course of their computations, the Chudnovskys identified a number of unexpected quirks peculiar to the computers they used for their calculations. Later, they continued their work on an IBM 3090 computer, using two different operating systems as a check.

One key feature of the Chudnovsky method is that computations of π are readily expandable. More digits can be added on demand, without having to restart the calculation from the beginning. That means this approach lends itself to group efforts. Much of the work of computing π to a given number of decimal places can readily be divided up among a large number of people, each one working independently on a small computer. The Chudnovskys envision a "π chain letter," with interested researchers, students, and hackers combining their efforts to do multibillion-digit calculations. Anyone with a computer, small or large, could add digits. The most serious bottleneck in such a scheme would be storing the intermediate and final results. But because the method is self-verifying, there would be no danger of an individual ruining the whole effort by making and passing on a mistake.

Computing π to a billion digits or more is important in the search for patterns among the digits of π. Although a billion digits aren't enough for doing a proper statistical analysis, preliminary results show some evidence for subtle analytic relationships among the numbers. But solid evidence for a pattern or any kind of order in the overall arrangement of π's digits is not yet at hand.

"In order to compute a number well, you have to know it intimately," David Chudnovsky says. At the same time, the techniques developed to compute a number, whether it's π or some other irrational, transcendental number such as *e*, brings you closer to a more complete picture of the mysterious relationships linking the mathematical fields of number theory and algebraic geometry.

Musical Numbers

There's a new mathematical ingredient in the sound of many performing artists and recording stars. It manifests itself in the form of clusters of panels hanging on the walls of recording studios, concert halls, nightclubs, and churches. Sculpted from wooden strips separated by thin aluminum dividers, each panel consists of an array of wells of equal width but different depths (see Figure 5.20). These panels, called reflection phase gratings, scatter sound waves. The result is a richer, livelier sound with an enhanced sense of space. Listeners claim that the panels seem to make the walls disappear. A small room takes on the air of a great hall.

The secret lies in the varying depths of a panel's wells. With depths based on specific sequences of numbers rooted in the mathematics of number theory, the wells scatter a broad range of frequencies evenly over a wide angle. The scientist who pioneered the ideas responsible for this development is Manfred Schroeder. About a decade ago, Schroeder and two collaborators undertook a major study of more than twenty famous European concert halls. One of their findings was that listeners like the sound of long, narrow halls better than that of wide halls. Perhaps the reason for this, Schroeder reasoned, is related to another finding that listeners prefer to hear somewhat different signals at each of their two ears.

In a wide hall, the first strong sound to arrive at a listener's ears, after sound traveling directly from the stage, is the reflection from the ceiling. Ceiling reflections produce very similar signals at each ear. But in narrow halls, the first reflections reach the

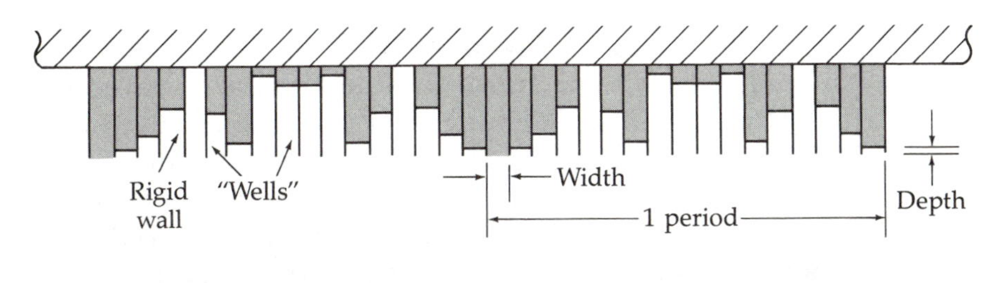

Figure 5.20 A sequence of wells of equal widths but different depths can scatter incident sound widely without loss, rather than reflect it like a mirror (*top*). Such reflection phase gratings permit the design of concert halls and studios in which sound is diffused throughout the chamber. Clusters of such gratings hang at the back of the control room at Tele-Image in Dallas, Texas (*bottom*).

listener from the left and right walls, and the two reflections are generally different.

This may be one reason why so many modern halls are acoustically unpopular. Economic constraints dictate the construction of wide halls to accommodate more seats, and modern air conditioning systems allow lower ceilings. To improve such halls, sound must be redirected from the ceiling toward the walls.

A flat surface can't do the job. It reflects sound in only one direction, according to the same rules that govern light reflecting from a mirror. The ceiling must have ups and downs that scatter sound with nearly equal energies going in all directions. Number theory can be used to determine the ideal depth of the notches, resulting in an acoustic grating that's analogous to diffraction gratings used to scatter light.

Although number theory suggests several answers, the most effective grating is based on quadratic-residue sequences. One example of such a sequence is based on the prime number 17. The first sequence member is the remainder, or residue, after the first number, 1, is squared and divided by 17. The answer is 1. Squaring all the numbers from 1 to 16, then dividing by 17 to find the residue, produces the sequence: 1, 4, 9, 16, 8, 2, 15, 13, 13, 15, 2, 8, 16, 9, 4, 1. For larger numbers, the pattern simply repeats.

This procedure is an example of modular arithmetic. Only remainders left over after division of one integer by another are saved. More formally, given any two integers, a and n, the remainder of a divided by n is written as $a(\mathrm{mod}\ n)$ and described as "a modulo n." In such an arithmetic system, 7(mod 5) is 2, 13(mod 5) is 3, and so on.

Finding the depth of a given grating well involves multiplying the appropriate number in the sequence by the longest wavelength for which the grating is designed to scatter sound efficiently and then dividing by a factor that depends on the well's numerical position. Mathematical analysis shows that for such an arrangement, the spectrum of energies scattered into different directions is essentially flat (see Figure 5.21, top), meaning that roughly equal amounts of energy go in all directions.

Why does number theory work so well? The answer is in the way waves cancel or reinforce each other, depending on whether

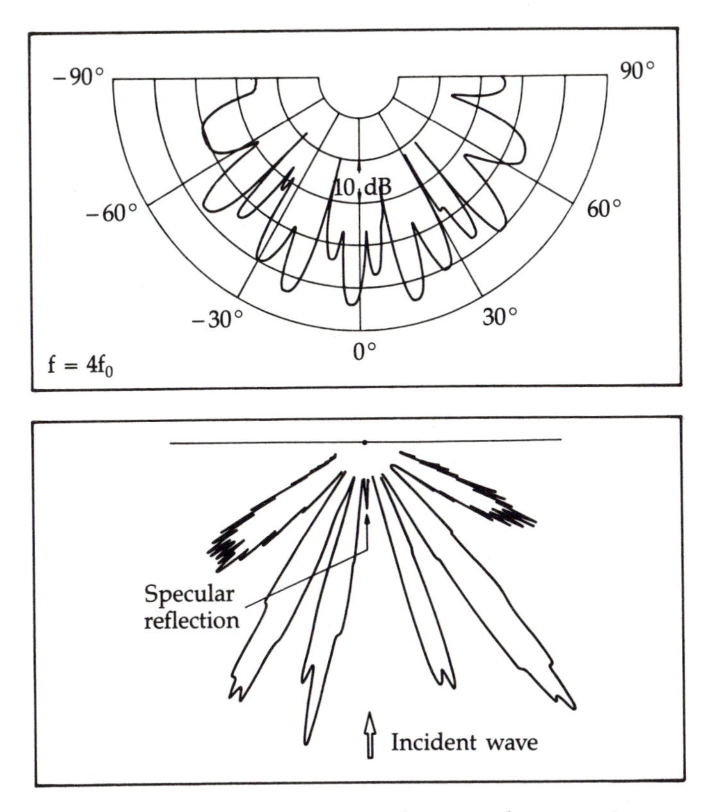

Figure 5.21 Sound scattered from a reflection phase grating spreads out over a wide angle (*top*). An array based on primitive roots scatters sound preferentially in certain directions (*bottom*).

the crest of one wave meets the trough or crest of another wave. For perfectly periodic waves, destructive interference occurs whenever one wave lags behind the other by half a wavelength or one-and-a-half wavelengths, or two-and-a-half wavelengths, or any other integer plus a half. In every case, the result is the same. Hence, in wave interference, it's not the path difference that determines the resulting pattern but the residue after dividing by the wavelength. Similarly, in modular arithmetic what counts is not the numerical value itself but the remainder after division by the modulus (17 in the example).

Reflection phase gratings based on quadratic residues are effective for scattering any type of wave. They spread out sound waves in concert halls, and electromagnetic radiation in the form of microwaves, radar signals, and light waves. Targets equipped with such gratings scatter radar signals well enough to give an enemy great difficulty in pinpointing the target. Similar gratings also scatter laser light effectively.

In some instances, it's useful to have strong scattering of sound waves or light waves to the sides and less scattering in directions corresponding to a direct reflection. That can be done by using reflection phase gratings based on *primitive roots*. For example, the prime number $N = 11$ has the primitive root $g = 2$, which means that the remainders after dividing 2^n by 11 assume all values 1, 2, . . . 10 exactly once as n goes from 1 to 10. In other words, the expression $(2^n) \pmod{11}$ generates a permutation of the integers from 1 to 10. In this example, $(2^n) \pmod{11} = 2, 4, 8, 5, 10, 9, 7, 3, 6, 1$, and so on, repeated periodically.

In a reflection grating, wells with depths based on primitive roots scatter sound waves so that components in all directions, except directly out from the surface, are equal in magnitude (see Figure 5.21, bottom). Thus, it's possible to construct reflection phase gratings that scatter incident energy (in the form of sound waves) relatively uniformly into different lateral angles, with little energy going into the specular, or mirror-reflecting, direction.

If increased scattering of waves is desirable not only to the sides but also forward and backward, then both the quadratic-residue and primitive-root designs can be implemented in two dimensions. Instead of long strips, the wells become an array of squares. Such gratings can be made out of wood blocks (see Figure 5.22). These two-dimensional diffusers may find important uses in the future for cutting down the amount of ambient noise. Noise scattered by such surfaces often becomes inaudible.

Reflection phase gratings now provide architectural acoustics designers with a new way of spreading sound around, both in space and in time. Designers have only three ingredients they can use to conjure up every imaginable type of acoustic environment,

Figure 5.22 Mounted on a concert-hall ceiling, a two-dimensional diffuser scatters sound both laterally and in the forward and backward direction. The spacers between individual wells are missing in this wood-block model of such a diffuser.

namely, absorption, reflection, and diffusion. Sound-absorbing surfaces made of foam or fiberglass and sound-reflecting surfaces, such as flat or curved panels, are widely used. Until reflection phase gratings came along, there really were no commercial sound-diffusive surfaces. For designers, it was like trying to type a paragraph without using, say, the letter "d."

At least one company now manufactures a variety of panels that can be clustered to create the right kind of sound patterns for control rooms, orchestra pits, or choir lofts. Conventional ensemble reflectors, which can often be seen standing behind groups of performers, are usually just curved or flat surfaces that reflect sound in only a few directions. An appropriately placed cluster of gratings, however, distributes the sound energy much more evenly. This pleases musicians because it allows them to hear their own instruments and gives them a better sense of how their playing fits in with that of others in the group.

Now that better digital recordings, electronic instruments, superior home playback systems, and new forms of music are available, demand is also increasing for improved acoustic surroundings for making and listening to recordings. Just like the performance of a symphony orchestra in a great hall, time cues in recorded material give us the perception of musical size, space, and depth. The preservation of these cues is vital to the overall enjoyment of any musical performance. Reflection phase gratings help to make three-dimensional sound coming from two speakers possible. By using gratings to diffuse sound and special sound-absorbing material near the speakers to minimize early reflections, it's feasible to create a listening environment in the home that allows a listener to experience a concert-hall ambience.

Number theory has long been considered the very paradigm of pure mathematics. Its application to acoustics and to other real-world problems would come as a surprise to many mathematicians more interested in pursuing the properties of integers. Even the term *integer*, Latin for "the untouched one," gives a sense a being above and beyond the concerns of everyday life.

Nevertheless, the theory of integers can provide unexpected answers to problems in the real world. Small integers and their ratios play a fundamental role in the construction of musical scales. Residue arithmetic plays an important part in digital communications, especially in providing cryptographic schemes to ensure protection of computer systems and data files. Integers and number theory also come up in electrical network problems, heat conduction effects, and game strategies. The days of modern life are truly numbered.

6

Hard Times

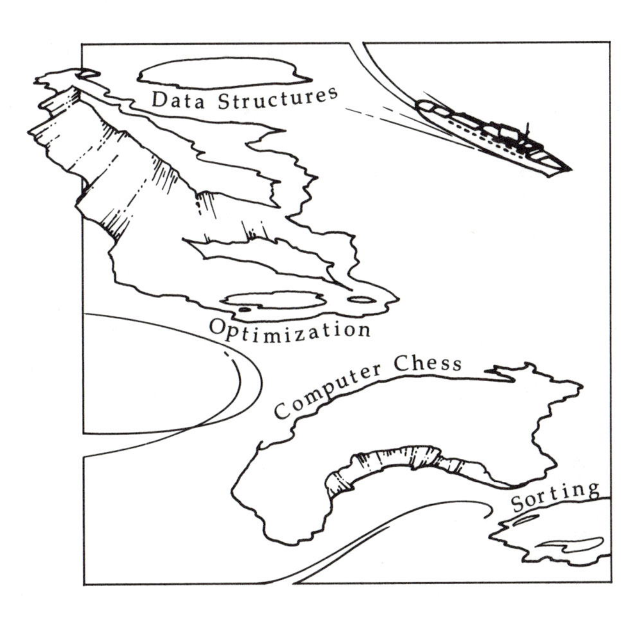

According to an ancient folktale from India, the Indian king Shirim wanted to reward his adviser Sissa Ben Dahir for inventing the game of chess. The king was so grateful, he offered a gold piece for each of the 64 squares on a chessboard. Sissa politely declined the offer and suggested an alternative reward. He said, "Majesty, give me a grain of wheat to place on the first square, and two grains of wheat to place on the second square, and four grains of wheat to place on the third, and eight grains of wheat to place on the fourth, and so, O King, let me cover each of the sixty-four squares of the board."

The king, astonished by such an apparently modest request, called for a bag of wheat and ordered his servants to count out the grains carefully, doubling the number of grains on each succeeding square of the chessboard. By the twelfth square, the king's servants had trouble fitting all the grains on a square, so they started making piles beside the chessboard. By the twentieth square, the bag was exhausted, and the king sent for more bags of grain. Finally, he gave up, still far from fulfilling his adviser's request. He could see there wasn't enough wheat in his kingdom, in India, or even the world to reach the sixty-fourth square.

How many grains of wheat had Sissa requested? The answer is $1 + 2 + 4 + 8 + \ldots$, which can be written as the sum of consecutive powers of 2: $2^0 + 2^1 + 2^2 + \ldots + 2^{63}$. That series adds up to $2^{64} - 1$, or 18,446,744,073,709,551,615. Assuming a ton of wheat contains 100 million grains, the quantity Sissa asked for amounts to roughly 200 billion tons of wheat. At the present rate of worldwide wheat production—less than 500 million tons per year—it would take more than four centuries to accumulate the needed quantity.

This tale demonstrates the danger of underestimating the power of simple mathematical operations that rapidly grow beyond reasonable bounds—a lesson not lost on computer scientists and mathematicians interested not only in solving mathematical problems but also in solving them as quickly as possible. In some instances, solutions to mathematical problems may require so many steps that no person would ever have the patience to wait for a computer to come up with an answer. It could easily take longer than a human lifetime. Identifying and characterizing such intrinsically "hard" problems is a major topic in computer science.

Math of a Salesman

Computer scientists often measure the difficulty of problems by the amount of time needed to solve them. That time depends on the nature of the problem, the type of computer, and the particular mathematical recipe, or algorithm, used to solve the problem. An algorithm is generally a set of instructions so complete and exact that no additional information or insight is necessary to accomplish the desired goal. Some algorithms are naturally faster than others, and some problems require methods that are necessarily time-consuming.

For example, one way to find the shortest path linking several points is to compute the lengths of all possible paths connecting all the points and then to select from the list the route representing the smallest length. That's easy to do when the number of points is small and the list is short, but every additional point greatly expands the number of computations required to come up with a list. The scheme falters because the total number of steps needed to solve a given problem may be so large that no computer can find an answer within a reasonable amount of time.

To get a handle on which problems are computationally time-consuming and which algorithms are fastest for solving a particular problem, computer scientists work with a theoretical notion known as *computational complexity*. Researchers try to measure how the number of steps needed to solve a given problem changes as the size of the problem increases. They use mathematical functions to express the difference between slow (inefficient) and fast (efficient) algorithms and between computationally easy and difficult problems.

To see how this notion works, consider how rapidly different functions grow. It's a game that can be played with a pocket calculator simply by substituting successive integers for n in the given expressions and calculating the answers.

	$n=1$	2	3	4	5	6	7	8	9
$f(n)=9$	9,	9,	9,	9,	9,	9,	9,	9,	9,...
$f(n)=2n$	2,	4,	6,	8,	10,	12,	14,	16,	18,...
$f(n)=n^2$	1,	4,	9,	16,	25,	36,	49,	64,	81,...
$f(n)=n^3$	1,	8,	27,	64,	125,	216,	343,	512,	729,...
$f(n)=2^n$	2,	4,	8,	16,	32,	64,	128,	256,	512,...
$f(n)=3^n$	3,	9,	27,	81,	243,	729,	2187,	6561,	19683,...

Clearly, some functions grow much faster than other functions (see Figure 6.1). Any constant function in which $f(n)$ equals a fixed value, will eventually be surpassed by any function proportional to n. It's also obvious that any function proportional to n^2 will outgrow a constant function or one proportional to n. Indeed, as the power of n increases, the function grows more rapidly.

Functions can also be expressed in terms of polynomials: expressions that combine different powers of a variable. Loosely speaking, such functions grow as fast as the expression's highest power. For example, the polynomial function $n^3 + 4n^2 - 6n + 2$ grows faster than any function with n^2 as its highest power.

Exponential functions, such as 2^n and 3^n, grow even faster than any polynomial function. No matter how slowly an exponential function starts out, it eventually surpasses any polynomial rival. And the higher the base of the exponential, the faster the function grows. Whereas each successive value of 2^n is 2 times larger than its predecessor, each value of 10^n is 10 times larger.

Thus, if a problem of size n can be solved n^2 steps, increasing n from 10 to 20 quadruples the number of steps. If the problem requires 2^n steps, then going from 10 to 20 multiplies the number of steps a thousandfold. Exponential growth clearly outpaces polynomial growth.

Function	Type	$n = 1$	$n = 3$	$n = 10$	$n = 30$	$n = 100$	$n = 300$	$n = 1{,}000$
$6n$	Linear	6	18	60	1.8×10^2	6×10^2	1.8×10^3	6×10^3
n^2	Quadratic	1	9	10^2	9×10^2	10^4	9×10^4	10^6
1.1^n	Exponential	1.1	1.33	2.59	17.5	13,781	2.62×10^{12}	2.47×10^{41}
2^n	Exponential	2	8	1,024	1.07×10^9	1.27×10^{30}	2.04×10^{90}	1.07×10^{301}
10^n	Exponential	10	10^3	10^{10}	10^{30}	10^{100}	10^{300}	$10^{1,000}$

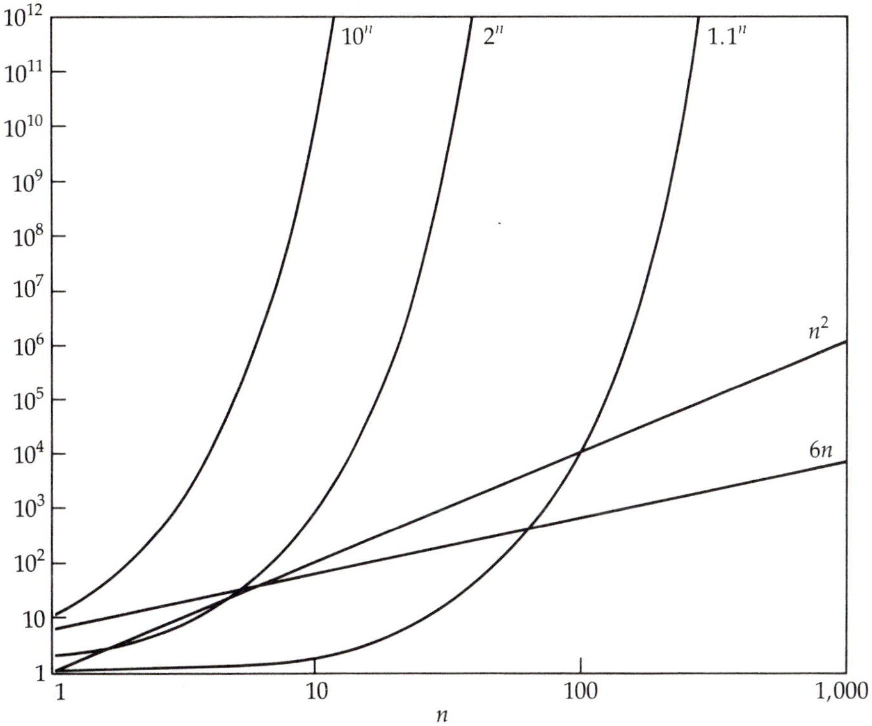

Figure 6.1 The efficiency of an algorithm depends on how its execution time grows as the size of the problem increases. The rate of growth is described by a mathematical function. Although a linear or polynomial function may exceed an exponential function for small values of n, beyond a certain n the exponential function is always greater.

Once computer scientists had means for estimating how efficiently algorithms work, they began to see a pattern. For certain problems, they could find algorithms in which the time or the number of steps required to solve the problems can be expressed as a polynomial function of the problem's size. Such problems are said to be easy to solve. For other problems, the best they could do was to find algorithms in which the time required to solve the problems is expressed as an exponential function of the problem's size. Such problems are difficult to solve not because there's no method available, but because all known methods take too long when the problem reaches a certain size.

One computationally easy problem is sorting a deck of 52 cards. Go through all the cards until you find the ace of spades, which you put aside, and then go through the cards again to find the deuce of spades, continuing in this fashion until all of the cards are in order. The worst that can possibly happen under such a sorting scheme is that the ace of spades is the last card in the unsorted deck, the deuce of spades the next-to-last card, the three of spades the one before that, and so on. If you started with n cards, at worst, you would have to take n^2 looks to sort the cards. The time or number of steps required to do the problem is a polynomial function of the problem's size.

The archetypal example of a "difficult" problem concerns a traveling salesman who must visit customers in a number of cities scattered across the country and then return home (see Figure 6.2). The problem is to find the shortest possible route visiting each city only once. Listing all possible routes, calculating the length of each one, and picking the shortest is one possible strategy for solving the problem. But for a large number of cities, such a brute-force procedure takes too long.

For instance, to find the shortest possible route to visit 10 cities, a computer would have to calculate 362,880 possibilities. For n cities, picking one city as the starting point, the number of possible choices for the next city is $(n - 1)$. Then you have $(n - 2)$ choices for the city after that, and so forth. The total number of possible paths is the product $(n - 1)(n - 2)(n - 3) \ldots (2)(1)$. As the number of cities grows, the number of possible paths skyrockets. Even the fastest computers available would require

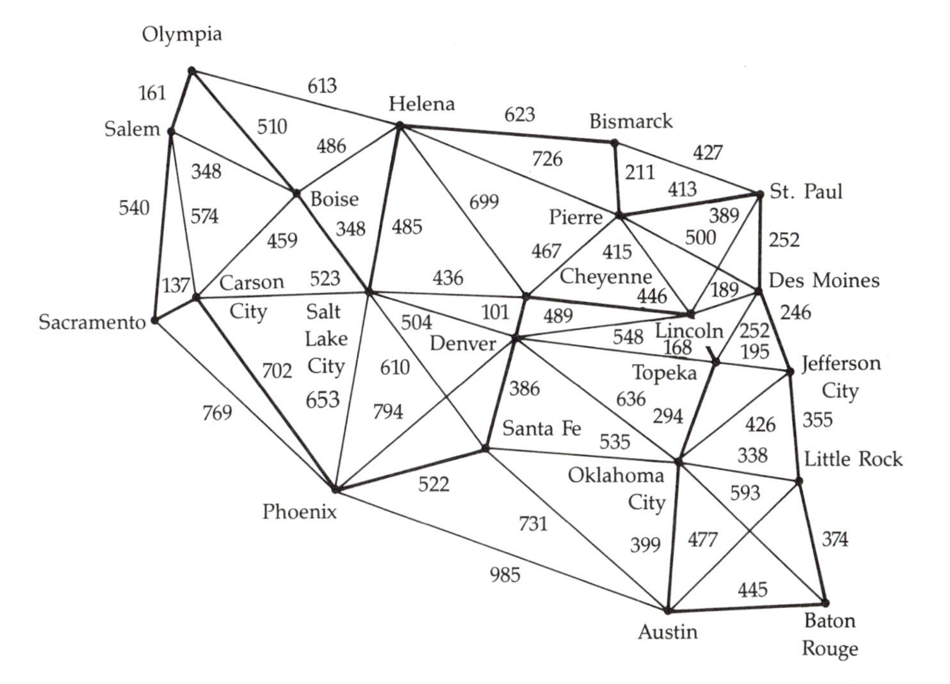

Figure 6.2 In the traveling salesman problem, a salesman wishes to visit all the cities in his territory, beginning and ending at his home city, while minimizing the distance traveled. The figure shows the state capitals of the states west of the Mississippi, with approximate road mileages between some of them. The route shown in bold has a total length of 8,119 miles. However, the shortest route has length only 8,117 miles. Can you find it?

years to handle the $(49 \times 48 \times 47 \times \ldots \times 3 \times 2 \times 1)$, or roughly 10^{62}, possible paths in a 50-city itinerary. The time required to solve this problem and many like it grows exponentially with the size of the problem.

Of course, individual instances of the salesman's problem may have simple solutions that work well under certain conditions. For example, first visiting all locations close to the starting point, then moving farther afield may work. But there's no guarantee that such a strategy works in every possible case. Indeed, proceeding from the origin to the nearest point, then to the nearest point to that, and so on, doesn't generally give the shortest path. In the end, the only way to be sure of getting the best answer is to try all the possibilities.

Problems similar to the traveling salesman problem crop up in many practical situations, such as designing networks to link an array of points, finding optimal routes under a variety of constraints, routing telephone calls, allocating resources, setting timetables and schedules, planning budgets, and packing objects into bins. Such problems come up in computer science, mathematics, physics, biology, economics, business, and the social sciences. Many children's puzzles and games fall into the same category.

In 1971, computer scientist Stephen Cook proved that a large group of these seemingly unrelated decision and search problems are all, in a mathematical sense, members of the same class. He showed there was a polynomial-time procedure for recasting any given problem of that class into any other problem of the same class. Thus, if it were possible to find an algorithm that solves any one problem in polynomial time, then all of the problems could be solved in polynomial time. As a group, the problems are labeled *NP-complete,* short for nondeterministic, polynomial-time-complete.

How many such problems are there? Cook and others have succeeded in classifying most naturally occurring decision and search problems as either easy or NP-complete. More than 2,000 fall into the "hard" category. The question of whether a polynomial-time algorithm can or will ever be found for solving these difficult problems is still unresolved. Proving the widespread conviction that no such algorithm exists for NP-complete problems is considered the fundamental open question in theoretical computer science. Finding such a proof is hard because it requires envisaging all possible algorithms for a given problem and demonstrating that each algorithm is inefficient. The real question is whether researchers just haven't been clever enough yet to figure out how to solve such problems efficiently.

Nonetheless, computer scientists can demonstrate that various problems are either easily solvable or provably intractable. The theory they have developed also shows that numerous different problems have essentially the same complexity. Computational complexity thus serves much the same function in computer science that the laws of thermodynamics play in the physical sciences. It establishes what's possible, distinguishing between what can and cannot be precisely computed.

Smart Guesswork

Demonstrating that a problem is hard, or NP-complete, provides strong circumstantial evidence that no efficient algorithm can be found for its solution. But it doesn't make the problem disappear. Other approaches must be developed to deal with the problem. Such a focus raises a host of challenging questions concerning standards for what constitutes an efficient algorithm and for what is an adequate though less-than-optimal result.

Suppose you have a certain number of identical bins and a collection of variously sized packages. Your task is to determine the smallest number of bins necessary to get all the packages in without overflowing any of the bins. Given that each package is smaller than a bin, the total size of the packages puts a lower limit on the number of bins needed.

The slowest and most inefficient way to proceed is to try all possible packings to see which one is the best. With two bins and two packages, you'd have to check 2^2, or 4, possible arrangements. For three packages and two bins, there would be 2^3 possibilities; for four packages, 2^4 possibilities. In general, there are N^n ways of stuffing n packages in N bins. Packing 75 parcels into 10 bins has at least as many possible solutions as the estimated number of atoms in the universe.

When such large numbers are involved, it's impossible to find the best answer. But often a simpler, faster method gives an answer that comes close enough for all practical purposes. One approach to the bin-packing problem, known as the first-fit method, is to arrange the packages in order of decreasing size. Then the bins are filled one by one by packing as many items as possible into each bin before going on to the next.

Given packages of sizes 10, 9, 8, 8, 6, 6, 5, 5, 3 for a total of 60 units, and bins each with a capacity of 20 units, the minimum number of bins required would be 3. Using the first-fit method of packing, 10 and 9 would go into bin 1; 8 and 8 into bin 2; 6, 6, and 5 into bin 3; and 5 and 3 into bin 4. The answer is 4 bins, but in this case, a three-bin solution does exist: $5 + 5 + 10$, $3 + 8 + 9$, and $6 + 6 + 8$ (see Figure 6.3).

Occasionally, the first-fit method does give the best answer. More often, it comes close. In general, for a large number of bins, the method arrives at a number of bins no more than 22 percent

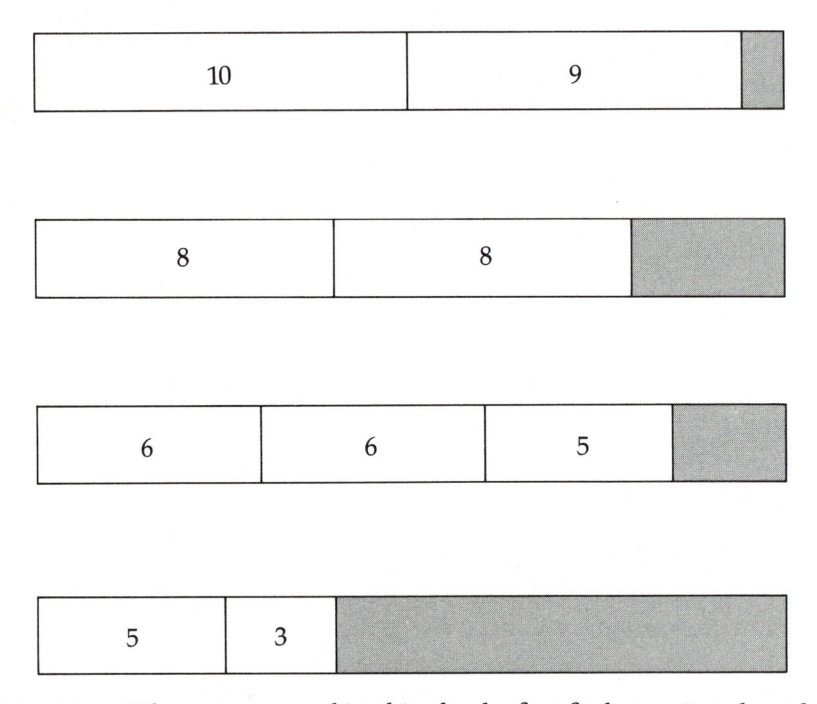

Figure 6.3 When you are packing bins by the first-fit decreasing algorithm, you consider the items in decreasing order of their sizes, and place each item in turn in the first bin that can accept it. In this example, there are a number of bins with a capacity of 20, and nine items ranging in size from 3 to 10.

greater than the optimal number. For many purposes, that's good enough. In daily life, a strategy of fitting the largest items in first is a useful hint to keep in mind when packing suitcases, boxes, or trunks. It's also a useful guide for a plumber who needs to cut various lengths of pipe from a number of standard-length pipes or for a paper manufacturer who must slice standard rolls of paper to provide customers with paper of a specified width.

The only way to solve many engineering and scientific problems, including the bin-packing problem, is to use strategies that take shortcuts to a satisfactory, though not necessarily optimal, solution. Often, the best approach is to start with a good guess, then apply mathematical techniques to improve it. The idea is to use methods that either restrict the number of possibilities to be tested to keep the search as small as possible or to follow a special route to reach a reasonably good answer quickly. Such strategies play a part in the most efficient algorithm known for getting a good answer to the traveling salesman problem. Developed by Shen Lin and Brian Kernahan, this algorithm speedily arrives at an approximate solution, usually only 1 or 2 percent away from the optimum. Researchers have developed many similar heuristic methods, such as the divide-and-conquer strategy (see Figure 6.4), and apply them regularly to a wide variety of real-world problems.

One novel, geometric shortcut for getting a good answer for the optimal-route problem is to use an "elastic net" to snare a reasonable route. The idea is to start with a small circle in the midst of a group of randomly scattered points (see Figure 6.5), which represent cities. Under the influence of "forces" exerted by the cities, the circle is gradually stretched in various directions until it forms an irregular loop that eventually passes near every city. It's as if the initial path is a circular elastic band, which becomes stretched until eventually it links all the cities by one of the shortest possible routes. Using computers that can simultaneously perform many operations in parallel, the elastic-net algorithm produces paths that are within 1 percent of the shortest known tours connecting a hundred randomly distributed cities.

The elastic-net algorithm originated in a mathematical model used to suggest how nerve cells in two different locations can be

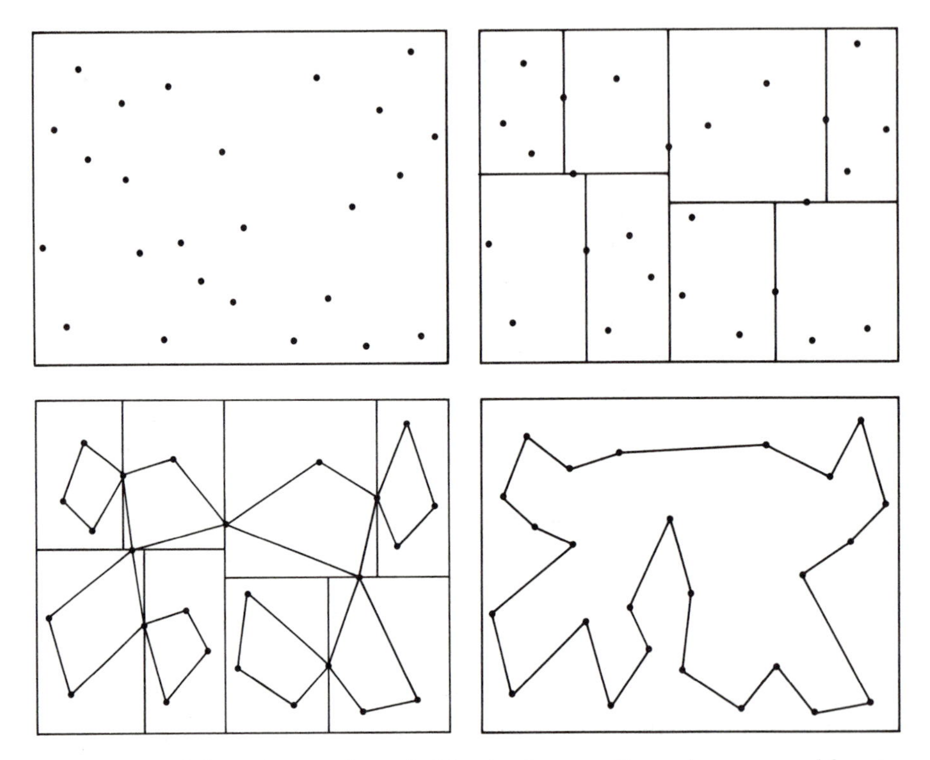

Figure 6.4 When the number of cities in the traveling salesman problem is extremely large, a simple divide-and-conquer algorithm will almost surely produce a tour whose length is very close to the length of an optimal tour. The algorithm starts by partitioning the region where the cities lie into rectangles, each of which contains a small number of cities. It then constructs an optimal tour through the cities in each rectangle. Finally, the algorithm combines these minitours and performs a kind of local surgery to eliminate redundant visits.

linked so that neighbors in one set of cells are also neighbors in a corresponding set of cells elsewhere. It provides a general method for matching one set of arbitrarily connected points to a second set of points located in a geometrical space of any dimension.

Another optimization algorithm, known as *simulated annealing,* is based on a mathematical model of the behavior of large

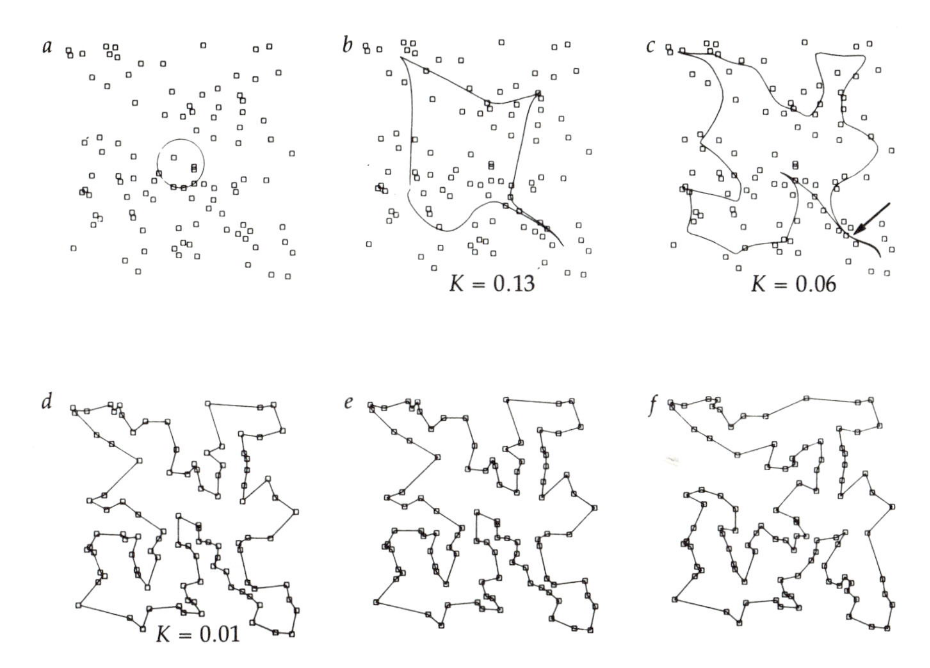

Figure 6.5 The elastic-net method for solving the traveling salesman problem for 100 cities randomly distributed within a square starts with a ring of a certain radius (a), then stretches it to encompass more and more cities (b, c, d). The final tour (e) has a total length of 7.78. The best answer yet found by any method for this particular case (f) has length 7.70.

collections of atoms or molecules under the influence of changes in temperature. This algorithm has become an important technique for drawing up timetables, designing microelectronic circuits, and processing images.

From long experience, metallurgists know that rapid cooling of a molten metal usually results in the formation of a solid made up of many small, irregular crystals, often riddled with defects. If the cooling is fast enough, the material may even end up in the form of a glass in which atoms fall into haphazard arrangements. Slow cooling, on the other hand, leads to the formation of large, well-formed crystals. Furthermore, metallurgists can transform a chunk of metal consisting of many small grains into one with only a few large crystals by raising the metal's temperature to nearly its melting point, then slowly cooling it. This process, known as annealing, allows the metal's atoms time to rearrange themselves into larger, more orderly arrays.

The same idea can be applied to optimization problems, for example, designing the complicated web of electrical connections on an integrated-circuit chip. The problem is to connect millions of microscopic transistors and other electronic devices while keeping the connecting wires as short as possible. The length of wire needed to connect a given arrangement of circuits is the quantity to be minimized. The computer randomly rearranges the chip's components, seeking a configuration with the shortest possible wiring, subject to any other constraints the designers have built into their simulation (such as keeping the wires far enough apart to avoid cross-talk or interference between wires).

The simulated-annealing algorithm requires four main ingredients: a concise description of the system's configuration, a random generator of moves or rearrangements of elements in a configuration, a way of determining how "good" a particular configuration is at any given time, and an annealing schedule specifying how quickly the system "cools down." During the process, the best rearrangements are those that lower the system's "energy." The trick is defining the equivalent of "temperature" and "energy" for a given problem.

The algorithm proceeds by taking the system in small steps from one configuration to another, and calculating the new configuration's "energy" at each step to see if its energy has gone down. To keep from getting trapped in a local minimum that is still far from the true minimum, the computer randomly allows occasional steps to a somewhat higher energy—a temporary change for the worse that often leads ultimately to an even better solution (see Figure 6.6).

When applied to the traveling salesman problem, the simulated annealing algorithm generates and tests random rearrangements of possible tours. The energy of a configuration corresponds to the length of the analogous trip. For 400 randomly distributed cities, the algorithm produces, within a reasonable time, a route only 0.1 percent longer than the theoretical optimum. It cuts down the number of steps needed to solve the

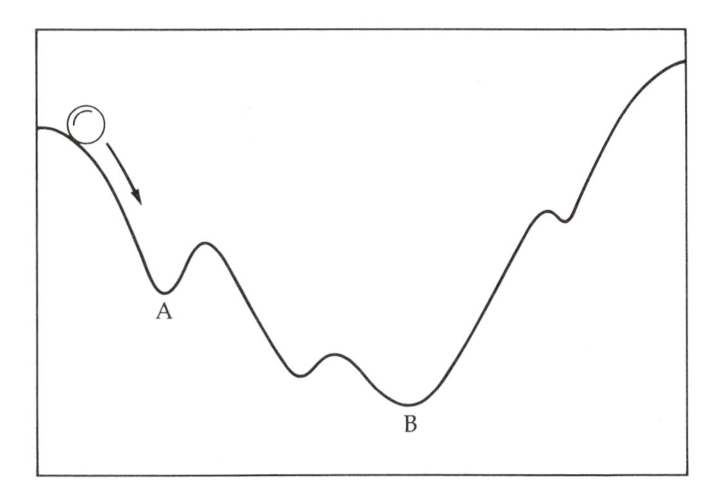

Figure 6.6 One way to imagine how simulated annealing works is to picture a hilly landscape, with a ball near one peak. Given a shake, the ball starts to roll, seeking the lowest possible position. However, sometimes the ball can get trapped in a hollow (A) that may still be far from the lowest possible position (B). But the simulated-annealing algorithm supplies random jolts that can jar the ball out of the trap and set it on its way.

problem from an astronomical 10^{800} steps required for testing every possible path to a more manageable 1 million or so trial-and-error steps.

Stimulated annealing turns out to be useful for a wide variety of scheduling and design problems. It's also handy in mineral prospecting for reconstructing the lay of the land from recorded echoes of seismic explosions. The method is particularly powerful for developing systems that allow machines to detect patterns and recognize objects (see Figure 6.7). In some instances, a computer can arrive almost unassisted at a solution that might have been thought to require the intervention of human intelligence.

Randomly stringing together a handful of resistors, capacitors, and transistors seems hardly the way to design and build a radio, but a random configuration is the starting point for a group of computer-based methods known as *genetic algorithms*. Whereas the simulated-annealing approach was suggested by sta-

Figure 6.7 These images of a group of objects (*left column*) have been corrupted by adding varying amounts of random noise. The simulated-annealing algorithm extracts and smooths the images hidden in the corrupted pictures to produce the results shown (*right column*).

tistical mechanics, genetic algorithms are rooted in the mechanics of natural selection and evolution. They represent a sophisticated kind of search that combines blind groping with precise bookkeeping.

The idea is to start with several random arrangements of components that each represent a complete but unorganized system. Most of these chance designs would fare very poorly, but some are bound to be better than others. The superior designs are then "mated" by combining parts of different arrangements to produce "offspring" with characteristics derived from both their "parents." From this second generation, the computer again selects the best or most efficient designs for further breeding, and rejects the rest. The process continues in this fashion until an acceptable design or solution to a specific problem emerges. Once the goal is clearly defined and the criteria for success are in hand, the computer itself picks its way in a trial-and-error fashion, recording and building on its best guesses and eventually producing a good answer.

Pioneered more than 25 years ago by computer scientist John Holland, genetic algorithms constitute a field of computer science inspired by biological models and strewn with biological terms. In essence, Holland links the question of how biological systems adapt to their environments with the problem of programming computers so they can learn and solve problems.

The genetic-algorithm approach to problem-solving has developed slowly. Only in recent years have researchers begun to appreciate and exploit the method's flexibility and versatility, especially for designing complex systems or finding near-optimal solutions to problems. Engineers are beginning to use genetic algorithms for such applications as designing integrated-circuit chips, scheduling work in a busy machine shop, operating gas-pipeline pumping stations, and recognizing patterns.

The first step in using a genetic algorithm to find a good design for, say, an integrated-circuit chip is to express all the chip's major components as a string of digits, or "chromosome," in which small groups of digits (somewhat analogous to genes on a chromosome) correspond to different components. The list of all possible chromosomes—different random arrangements of the components—would correspond to all the available design choices. In addition, the designer must specify a way of numerically evaluating the efficiency or quality of any given design.

A typical run starts with a small selection of chromosomes. In each succeeding generation, the method creates a new set of chromosomes using the best pieces of the previous generation—a kind of survival of the fittest. Such an approach turns out to be surprisingly efficient, partly because it builds on previous "answers," making it unnecessary to search the entire field of possible designs. Each trial after the first becomes less and less random. Once it eliminates bad parent strings, the method automatically eliminates their offspring, the offspring of their offspring, and so on. That allows a genetic algorithm to zero in rapidly on a good solution.

To keep the algorithm from getting stuck at a significantly less-than-optimal answer (perhaps because of an unfortunate choice of starting configurations), genetic algorithms also include a "mutation" operation. Used sparingly, this operation randomly changes a digit in the string of digits that make up a chromosome. However, the mutation operation plays a minor role compared with the key role played by the process of splitting chromosomes and then piecing together various fragments (see Figure 6.8). Recombination rather than mutation is the driving force.

That particular insight has encouraged a few biologists to take a closer look at the comparative importance of mutation and

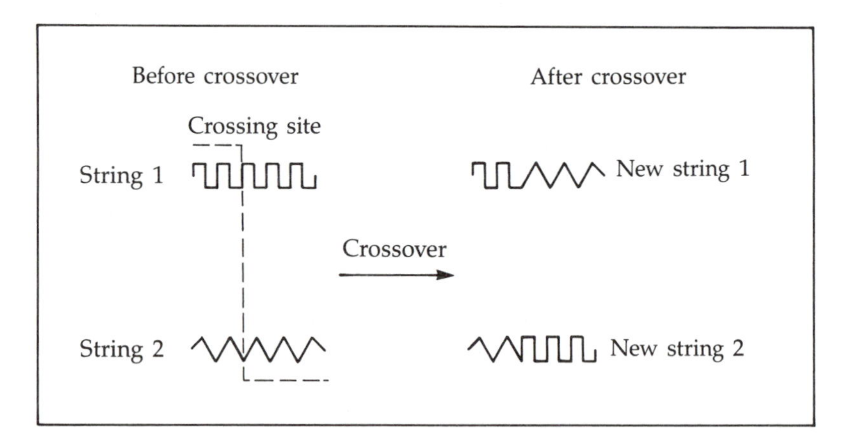

Figure 6.8 The most important phase of a genetic algorithm is the cut-and-splice (crossover) operation. In the simple crossover shown, two strings are cut at a site chosen at random, then the pieces are exchanged to create hybrid offspring.

genetic recombination in biological evolution. Evolution governed strictly by random mutation seems too slow a process to account for the present-day diversity of life forms. According to Holland's model, mutation's only real function may be to restart an evolutionary process that has momentarily stalled.

Simple genetic algorithms can handle such basic engineering tasks as designing a minimum-weight truss—a framework of struts meant to support a roof or some other structure. Although engineers have many methods for working out such problems, a genetic-algorithm approach proves astonishingly quick in narrowing the choices for constructing, say, a 10-member truss. In such a computer program, a string of 10 numbers (a chromosome), with each number describing the diameter of a particular strut, represents the 10-member truss. A computer initially generates from 50 to 100 chromosomes and then proceeds to weed out the weaklings while reproducing, splitting, and recombining the superior combinations. It takes about 40 generations to arrive at a near-optimal design.

The speed with which a genetic algorithm can find solutions allows it to "learn," to adapt to changes in its environment in a way that conventional expert systems, which rely on a complicated web of rules to specify responses, can't match. In general, expert systems embody attempts to put knowledge in a form that allows machines to mimic human reasoning within a limited context. The trouble is that conventional expert systems are too rigid, or "brittle." The minute such a system has to handle anything outside its domain, its answers become nonsense.

By showing that a search can proceed efficiently without any prior knowledge of a problem's structure, the genetic-algorithm approach provides a possible answer to the question of how machines can learn (see Figure 6.9). Various systems based on genetic algorithms have managed to learn simple tasks such as negotiating a maze and playing draw poker. On a more practical level, one genetic-algorithm-based program copes with the ever-changing demands on a gas-pipeline pumping station. Faced with fluctuating input because of pump breakdowns, line leaks, and daily and seasonal pressure changes, the operator of the system must learn how to adjust the pumping equipment to maintain a

	Trial strings	Scores
A	0 1 0 1 0 1	1
B	1 1 1 1 0 1	1
C	0 1 1 0 1 1	3
D	1 0 1 1 0 0	3

	Multiplication
C	0 1 1 0 1 1
D	1 0 1 1 0 0
C	0 1 1 0 1 1
D	1 0 1 1 0 0

	Division
C	0 1 : 1 0 1 1
D	1 0 : 1 1 0 0
C	0 1 1 0 : 1 1
D	1 0 1 1 : 0 0

	Crossover offspring	Scores
E	0 1 : 1 1 0 0	3
F	1 0 : 1 0 1 1	3
G	0 1 1 0 : 0 0	4
H	1 0 1 1 : 1 1	3

Figure 6.9 A simple game illustrates how a genetic algorithm would work. Your opponent secretly writes down a string of six digits, each either a zero or a one. You, in turn, suggest strings, and your opponent scores them, the score being the number of digits in a trial string that match digits in the secret string. The object is to match all the digits in the secret string in as few moves as possible. The procedure begins with four random trial strings (A, B, C, D). The low scorers (A, B) are deleted, and strings C and D are duplicated so that four strings are still in play. These "parents" are then mated by cutting each at an arbitrary point and splicing the first part of one (C) to the second part of the other (D). The new strings are scored, and the entire operation is repeated with the best offspring, until a perfect score is reached.

steady output. The computer program for running a pumping station finds a combination of pumps and pressures that achieves the desired output. As the input changes, the algorithm automatically searches for a new combination that keeps the flow as steady as possible and minimizes the amount of natural gas used to run the pumps. The system essentially trains itself.

Civil engineer David Goldberg describes the potential of genetic algorithms in the following words: "Natural evolution has found all kinds of interesting organisms to fill many different niches. Genetic algorithms are broad in the same way. Although they may not be tailored for solving a particular problem in the best possible way, they work well for many different problems."

A Different Sort

Microseconds matter. Every extra step that a computer takes to find a bit of data or to do a simple operation takes time. Repeated thousands of times, the extra steps add up to long delays in computer processing that may stretch into minutes or hours. Sorting and searching, in particular, take a lot of time. In fact, most computers do relatively little computing. They serve as gigantic filing systems where your savings account or tax record jostles with millions of others. Finding your account without examining everything else in the data bank depends on keeping the data in some kind of order, and that means sorting. Every time something is changed, added, or removed, there's more sorting.

Many computer scientists and applied mathematicians have thought long and hard about ways to make computer searches as simple and efficient as possible. This has led to the invention of a variety of methods for organizing information into *data structures*. In general, a data structure—whether in the form of a list, table, array, or tree—specifies the items involved (perhaps a set of names, telephone numbers, or key words) and what can be done to those items. One example of a simple data structure is a list of names and telephone numbers constructed so that specific telephone numbers can be found, new numbers added, and old numbers deleted.

In the case of a telephone directory, the most naive way of programming a computer to find a specific number is to ask it to start at the top of a list and keep on going, checking each entry one by one, until it happens on the requested item. If frequently used numbers appear near the end of a list, the computer could spend an unnecessarily long time searching for the numbers. One way of speeding up a search—either to find a given item or to insert a new item in an appropriate slot—is to sort the list into a particular order before using it. But what's the most efficient way of sorting a list?

One of the oldest ways of sorting known to computer science is the *bubble sort*. Suppose you have laid down a row of 13 cards of the same suit from a deck and you want to put the cards in order from ace (lowest) to king (highest). Scan the row from left to right. Whenever the left member of a pair is higher than the right, exchange the cards. From the moment you first encounter the king, every subsequent swap involves the king. The king ends up at the right hand end of the row. Starting again brings the queen next to the king, and so on, until all 13 cards "bubble" into place one by one (see Figure 6.10).

Although time-consuming, the bubble sort is simple and depends only on local actions: knowing which number of a pair is greater and, if necessary, switching the positions of the numbers.

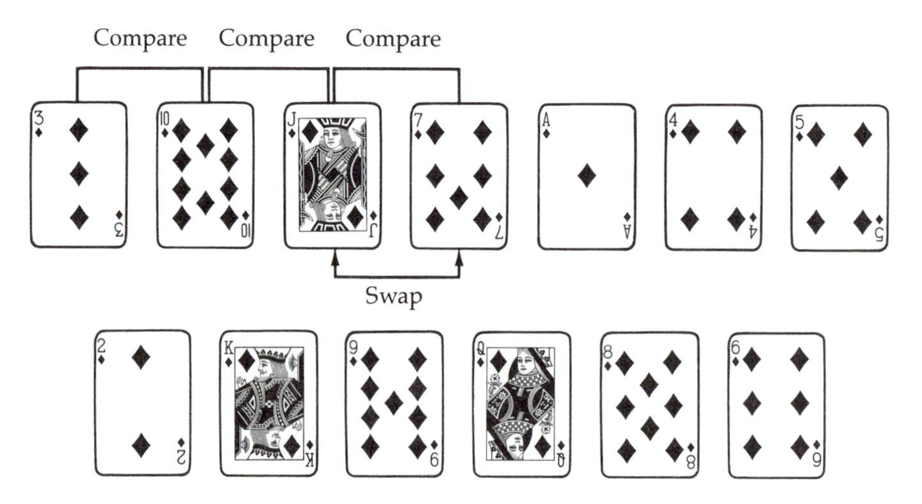

Figure 6.10 A bubble sort uses a series of comparisons between neighboring objects (in this case, playing cards of the same suit). If the higher-valued object isn't on the right, the objects are swapped.

It's the kind of simple scheme that anyone can do without having to understand what is going on. The same technique allows strangers to organize themselves into an ordered line. Suppose fans of a rock group were all given numbers to determine their places in line to buy tickets for a concert. Even if the fans wandered off while waiting for the box office to open, they could still easily get back in line. To get in order, each person would look at the number of the person behind him or her. If the number were higher, the two people would exchange places. A flurry of such exchanges would put everyone in their rightful place.

The trouble with the bubble sort is that as the number of items grows, the number of steps required increases dramatically. Its fatal flaw is that a high-numbered card can get to where it belongs only by single-step exchanges. A card player, however, knows to move a king to the far right in one swoop, which suggests an alternative strategy, the *selection sort* (see Figure 6.11). The selection sort involves less swapping and more comparing. The idea is to scan the row for the highest card, then swap this

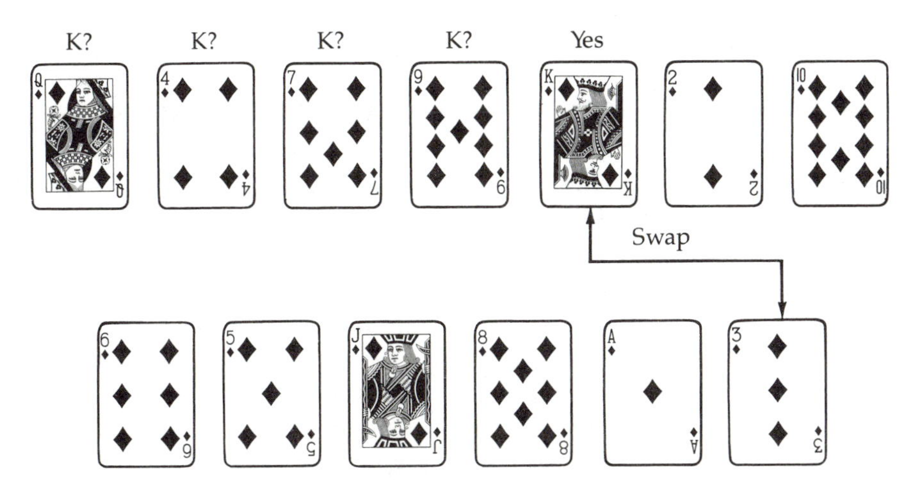

Figure 6.11 In a selection sort, the first scan looks for the highest-value object (the king in this grouping of cards of the same suit), then swaps it with the object at the extreme right. The second scan looks for the object with the second-highest value, then swaps it with the second card from the right, and so on.

card with the card to the far right. Because swapping takes more computer time than comparing, the selection sort is usually faster than the bubble sort.

However, the bubble sort can be the faster method when the items are already mostly sorted, which actually happens quite frequently, especially when only one or two newcomers need to be inserted into an existing arrangement or list.

Another scheme is familiar to anyone who has ever organized a hand for playing bridge or sorted a month's worth of canceled checks into order by date. The algorithm is to take each item one at a time and move it into the correct place with respect to all the other items already sorted. In other words, the second item is placed before or after the first. The third item is compared with each of the two items already sorted to determine its place, then inserted into a place created by shifting neighbors each one place rightward, and so forth. Sorting five items would require up to 25 comparisons to achieve a sorted list; n items would require n^2 comparisons. But for more than 50 items, the process quickly becomes cumbersome and no longer worth using.

These examples are only a few of the ingenious sorting methods computer scientists have developed. Each one has its own advantages and disadvantages. Some methods work better with lists that are nearly random to start with, while others work best with lists already almost sorted. What about an "omnisort"? Is there a *best* way to sort files, records, or data? So far, no one has either suggested how it might work or proved that it doesn't exist.

One way to save time is to combine sorting with searching, perhaps by allowing items within a list to reorganize themselves according to how often they are called upon. Such a strategy gives rise to a *self-adjusting data structure*. For example, each accessed item can be automatically moved by swapping places with its predecessors so that it gets closer to the front of the list. Because the most frequently accessed items will end up at the front, the computer will find them more quickly. Although the move-to-front operation itself takes some time, the hope is that over the long run, there will be a net saving. Alternatively, a single exchange, in which the accessed item is swapped only with its immediate predecessor, may be good enough to decrease computer search times.

The general aim is to make small changes that improve the organization of a data structure if such moves are applied repeatedly. Self-adjusting data structures have several advantages. They're very simple, which makes them easy to program. They also adjust to fit input data or changing patterns of requests. Although other, more efficient algorithms for searches have been invented, most turn out to be so hard to program that people are not willing to take the time to do it. With a fixed data structure, if the data coming in don't fit the model of what the data are supposed to look like, the system slows down considerably.

But how efficient are self-adjusting schemes in general? Although these data structures are extremely simple to program, they're very difficult to analyze because their structures are constantly changing, and mathematical proofs establishing the limits of their behavior are hard to come by. So far, both theoretical results and experiments performed on real data indicate that the move-to-front scheme works efficiently in a wide variety of situations. For instance, it works well when tried on English-language text and on the names of variables in computer programs written in languages such as Pascal.

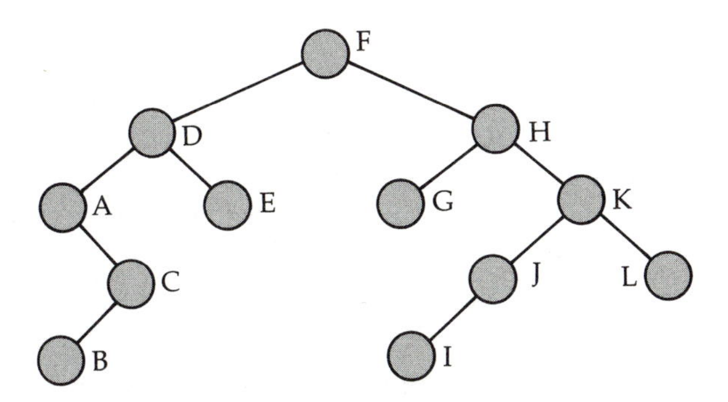

Figure 6.12 The items in a binary search tree are arranged so that all items in the left subtree are less than the item in the root, and all items in the right subtree are greater. In this example, the items are ordered alphabetically.

Computer scientists Robert Tarjan and Daniel Sleator have applied the same ideas to binary search trees, in which items are organized so that they are linked in a definite pattern (see Figure 6.12). The starting point (or root) branches into two nodes. These nodes, in turn, each branch to form two more nodes, and so on. In such a tree all items in the left branch (or subtree) are less than the item in the root, and all items in the right branch are greater. A tree organized in this way supports fast retrieval of ordered items stored at its nodes. The deeper the item, the longer the search takes because more nodes need to be checked.

The question is whether there is an equivalent of the move-to-front method for these binary search trees. In other words, if the computer often needs information stored at particular nodes, is there a way to move those items close to the root and so rearrange the tree to make searches more efficient? The answer supplied by Sleator and Tarjan is to use a move called a "rotation," which shifts the connections between the nodes in a specific way (see Figure 6.13, top). A combination of rotations to give a sequence of steps called a *splay* operation moves the retrieved node to the root of the tree and, as a bonus, also approximately halves the distance from the root of the tree (where the computer would start its search) to all of the nodes visited during the search. That shift makes all nodes visited during the retrieval much easier to find on subsequent searches (see Figure 6.13, bottom).

Thus, whenever an item is retrieved, certain adjustments to the tree take place along the access path leading to the appropriate node. Although restructuring the tree takes time, the overall procedure is very fast in many situations for a long sequence of retrievals. This is especially true when the pattern of requests is nonuniform or somewhat unpredictable. Restructuring the tree also makes operations such as insertions and deletions simple and quick.

Self-adjusting search trees turn out to be very useful for allocating space within a computer's memory. When working with large, shared computers, users are continually asking for blocks of storage space of a certain size. As the number of users and their needs change, the computer's operating system must have a way of allocating new requests for space. In the simplest approach, the

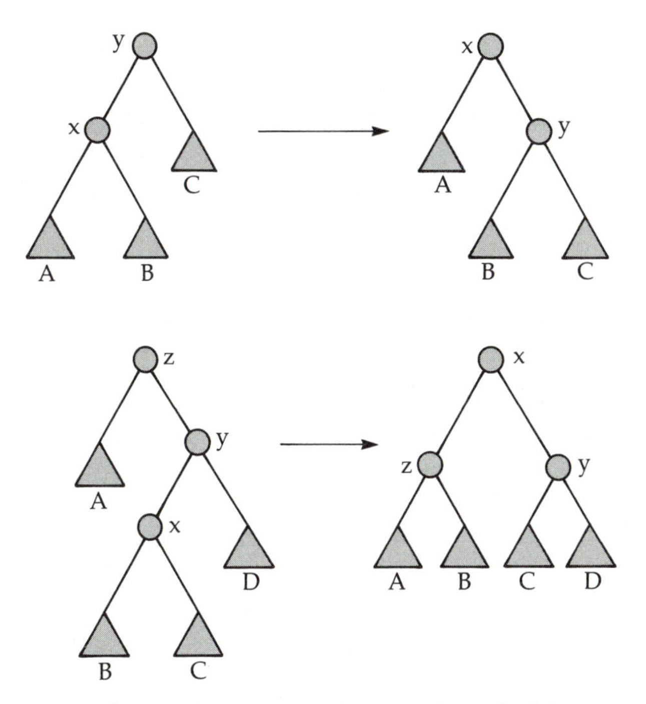

Figure 6.13 A single rotation rearranges a portion of a binary tree (*top*). A splay step, involving several rotations, brings node *x* to the root of a simple binary tree (*bottom*).

computer scans, from front to back, a list of all the available memory blocks until it finds a "hole," or free space large enough to accommodate the request. The problem is that the front of the memory gets chopped up, usually leaving only tiny chunks free. The computer has to scan over an increasing number of these small holes until it finds one large enough to use. The memory-allocation routine itself starts to slow down.

Using a self-adjusting binary search tree to represent the holes significantly speeds up the allocation process. Each node of the tree corresponds to a block of memory. In fact, information about the tree itself can be temporarily stored in the free blocks. The tree structure sits in the unused space, and as the unused space fills up, the tree gets smaller.

Researchers are now looking at more complicated search trees and at arrays that contain several bits of data in each location. They are also exploring other situations in which the notion of self-adjustment may apply.

This type of research plays an important role in the design and analysis of efficient computer algorithms. The goal in this field is to devise problem-solving methods that are as fast and use as little storage as possible. The idea is to perform theoretical analyses that give bounds on a given algorithm's use of computer time and space.

"Designing for theoretical efficiency requires a concentration on the important aspects of a problem, so as to avoid redundant computations and to design data structures that exactly represent the information needed to solve the problem," Tarjan says. "If this approach is successful, the result is not only an efficient algorithm, but a collection of insights and methods extracted from the design process that can be transferred to other problems."

In the end, the design of efficient algorithms depends on picking appropriate data structures. Such structures can be complicated and sometimes take years to analyze and understand. "But a good idea has a way of eventually becoming simpler and of providing solutions to problems other than those for which it was intended," Tarjan continues. "Just as in mathematics, there are rich and deep connections among the apparently diverse parts of computer science."

Playing Bit by Bit

Human players still rule the game of chess. World chess champion Gary Kasparov proved that in October 1989, when he easily crushed his electronic challenger, Deep Thought. But how long will human players maintain this advantage over chess-playing computers?

That such a match took place at all provides a striking demonstration of how far computer chess has advanced in recent years. Only a decade ago, an expert player could defeat any computer program in existence. Now all but the 100 or so strongest players in the world would lose to today's best chess computers. However, Kasparov's decisive wins in both games of his match against Deep Thought vividly show how much farther researchers have yet to go in designing a champion chess machine.

Put together by a team of five graduate students at Carnegie-Mellon University, Deep Thought played its first game in May 1988. Since then, the custom-built computer has amassed a remarkable record, including several wins over players ranked as grandmasters and only a handful of losses. Early in 1989, Deep Thought became the first machine to earn a chess rating that put it into the top category of all chess players. It attained a rating of 2550, which gives it the rank of grandmaster. World champion Gary Kasparov's rating is close to 2800 (see Figure 6.14). In May 1989, Deep Thought also won the world computer chess championship, confirming its status as the world's best chess machine.

Computer scientists and other researchers have been thinking about machine-based chess for a long time. In 1950, Claude E. Shannon, pioneer of information theory, explained why he thought chess a good challenge for researchers: "The problem is sharply defined, both in the allowed operations (the moves of chess) and in the ultimate goal (checkmate). It is neither so simple as to be trivial nor too difficult for satisfactory solution."

Apart from the sheer intellectual challenge of the task of programming a computer to play chess, chess turns out to be useful for checking out new programming concepts that may eventually prove useful in other applications. In a sense, chess programs have the same role in developing machine intelligence that fruit flies, which reproduce quickly and go through many generations in a short time, have in genetic research.

Specific techniques developed in programming a computer to play chess have been used in writing programs for solving other types of problems, particularly those involving a search among many alternate pathways, as in a telephone-switching system or an electric-power grid. There is also the possibility that methods for structural pattern recognition on a chessboard will eventually help computer-operated robots to see. At a deeper level, chess programs could lead to important insights into how the brain works—how it analyzes patterns and quickly abstracts what is important from what is not so important.

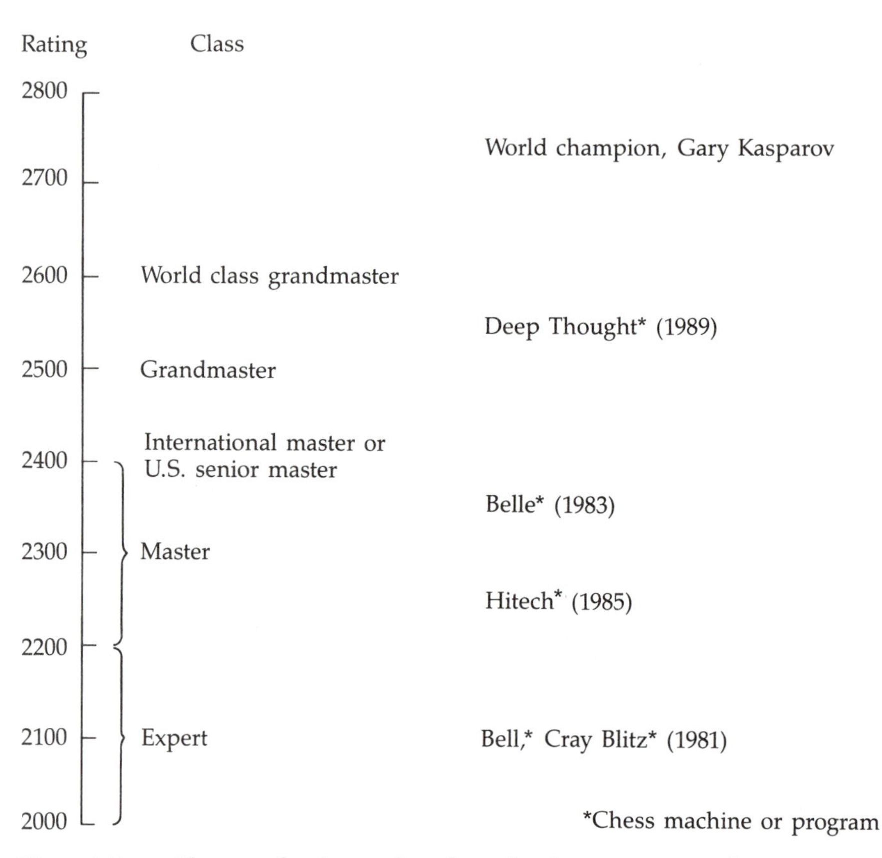

Figure 6.14 Chess enthusiasts who often play in tournaments have a rating that ranges from 1000 to around 2800. These ratings are computed by formulas that predict the score a player should make in a tournament or match, depending on a player's own previous rating and his or her opponents' ratings. If a player does better than expected, the rating goes up; after a worse score, the rating drops. Chess computers have steadily improved their standing in recent years. Deep Thought, in particular, now plays at the grandmaster level.

Deep Thought's performance is an apt illustration of how far an intelligent, speedy search can go toward mimicking human capabilities. All of today's chess programs and machines are fundamentally alike in that they depend largely on a systematic, exhaustive search. A computer looks ahead from its current position along a branching tree of possibilities. The program assigns a value to the end of each branch according to its strategic strength or weakness. These values are then compared, and the computer finally decides how to move. Deep Thought can scan nearly a million such positions per second.

But Deep Thought and other chess computer's don't play their games simply by working out every possible move and all its consequences for a given position. There are just too many possibilities for each one to be considered within the time limits that govern chess games.

In principle, any system capable of evaluating 10^{125} positions and having a sufficiently large memory could determine the outcome of any chess game even before white makes its first play. But because the universe as a whole contains only 10^{80} particles, it's highly unlikely any such system could be developed. Even evaluating the opening moves of a chess game escalates rapidly into an impossible search. A player starts with a choice of about thirty different moves, and all the possible responses results in a total of 900 different opening positions. That total branches rapidly into millions of possible positions after only a few more moves.

Whether playing against a human opponent or another chess machine, a chess program, like a superb racing car, is fine-tuned to run as fast as possible. Some programs are designed to take advantage of special features built into the world's fastest computers. Other electronic chess prodigies are high-speed machines in which chess instructions are wired into circuits or etched onto chips. Many have extensive libraries of standard opening moves and other aids for saving time during a game.

Typical computer chess programs are fast enough to look ahead about four moves. Various pruning tricks shorten the search by taking out many useless and obviously silly moves. Another important shortcut for the program is to assume that its opponent is smart and will generally take one of the best moves available. The program need not search much beyond the point at which it identifies a particularly strong move its opponent can make. Then it can concentrate on responding to such a possible move (see Figure 6.15). Deep Thought, for instance, incorporates an innovative search procedure known as "singular extension,"

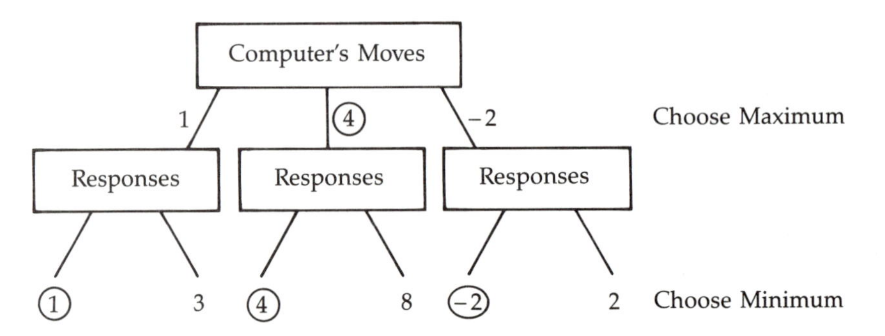

Figure 6.15 A sequence of moves and countermoves forms a tree structure, with nodes in the tree representing board positions and branches representing moves. Chess programs stop their searches at some point in such a tree and assign a score to that position. If the computer is winning, the score is positive; if its opponent is winning, the score is negative. The value of a move for the computer is the minimum of the values of the countermoves available to its opponent. The value for a countermove for its opponent is the maximum of the values of the counter-countermoves available to the computer. In the so-called minimax strategy, the computer scores moves by alternately choosing the maximum and minimum value of the next move along.

which allows the machine to probe as many as 30 moves ahead along promising tracks instead of staying with a general search.

The chief weakness of a scheme relying mainly on a fast search is that the program is oblivious to problems and traps that lie beyond the reach of its search. That's the kind of weakness Kasparov exploited. "The computer is too simple, too straight, too primitive," Kasparov said before his first game against Deep Thought. "If you know a computer well, you can anticipate its moves."

Indeed, Kasparov played in an uncharacteristically deliberate fashion. Avoiding the daring moves and slashing attacks that typify his most famous games, he played more like a stern schoolmaster setting a carefully crafted test designed to put a precocious but inexperienced student in his place. Kasparov deftly maneuvered Deep Thought into positions that forced the computer to make the wrong moves. Unable to search far enough ahead to gauge the effect of key moves and distracted from mounting a concerted attack, Deep Thought readily fell into the traps Kasparov set.

When a spectator watches a game without knowing which player is the human and which one the computer, certain characteristics sometimes give away a computer's play. For example, a chess computer may make an obviously stupid move—like shifting the king one square left or right on the back row. It may do that every once in a while for almost no reason whatsoever except that it can't find anything better to do. Moreover, a chess-playing computer often tends to favor gaining small immediate advantages over eventual large gains (see Figure 6.16).

On the other hand, human players may have something to learn from a computer's somewhat cumbersome and cautious play. Although its games against Kasparov revealed considerable shortcomings in Deep Thought's play at the beginning of a game, they also showed the computer's strength in making the best of a bad situation. Faced with an unfavorable position so early in a game, a human player would either give up or try something drastic. The computer, however, can stave off defeat for a long time, making the best possible moves available while waiting for its opponent to commit a mistake. Unfortunately for Deep Thought, Kasparov played flawlessly.

In general, games in which a human plays a computer offer more surprises than games between chess-playing computers. Because every computer is using basically the same algorithm, the ones that can search most deeply almost always win. Although searching a little deeper is a great advantage against another program, it isn't always much of an advantage against a human being.

A chess computer's speed and patience often make up for its other limitations. Good chess programs play consistently and don't get tired or suffer from headaches, excesses of passion, and bouts of nervousness. One computer-chess afficiando describes a chess-playing computer as a shark. "It's not very bright, but once it gets a taste of blood, it gets right in there and goes munch, munch, crunch," he says, adding, "A player can build up a nice attack disguised behind layers of pawns, and the computer may not suspect what's about to happen. But if you allow any slight

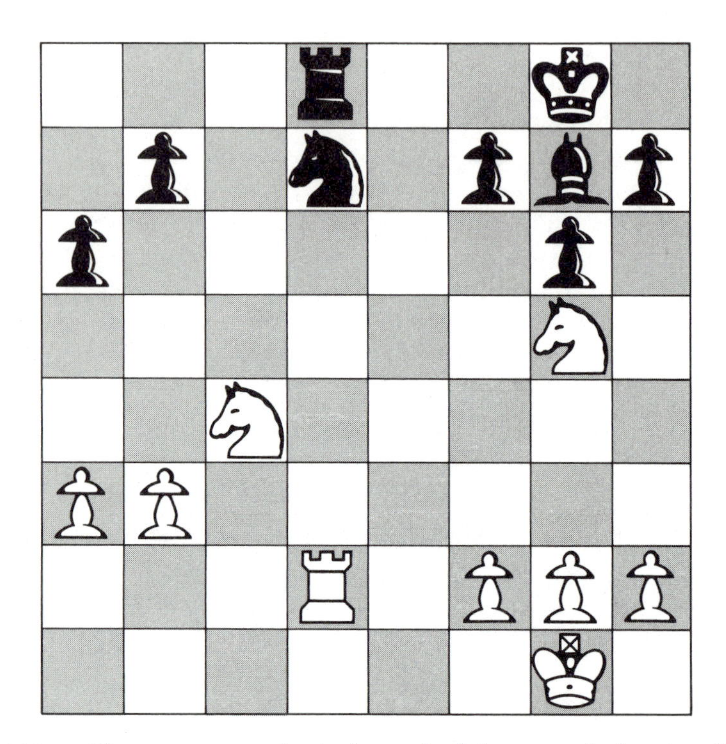

Figure 6.16 Chess programs don't always look far enough ahead to discern what is happening. In this example of the "horizon effect," if black continually attacks white pieces, white doesn't have time to continue attacking the knight to win it. A human understands that delaying the attacks does not eliminate them, but a computer, by forcing the attacks out of its search, thinks that it has totally avoided the problem.

chink in the armor, you may suddenly find this thing coming after you, with all your nicely laid plans in ruins."

Indeed, chess computers can drag their opponents into bizarre situations, and they force human player to think more deeply about the nature of the game. By contesting a machine, which plays chess somewhat differently, players gain valuable insights into their own weaknesses, foibles, and style of play. Computer scientist Ken Thompson of Bell Labs, one of the creators of the chess machine Belle, once commented: "I play chess, pit myself against the computer, as a barometer of where I am. One gets a fresh point of view. The computers are always original. They're not enslaved by what's been done in the past."

In their efforts to build and program chess-playing computers, researchers are also learning new things about the game of chess itself. Systematic studies of how to end chess games when both players have only two or three pieces on the board have already shown that much conventional wisdom is flawed. Chess computers running through published games are also finding errors in many books that describe standard and innovative game openings.

Brute-force searches by computers, put to the task of solving all possible endgame positions with five or fewer pieces, have turned up a number of such surprises. The computer begins by listing every possible checkmate, then works backward to find every possible predecessor for a given position. Using some ingenious logic, it eventually comes up with a list of all the positions in which one side can force a win. This takes some doing because five pieces can be arranged in hundreds of millions of different chess positions.

What the exercise makes clear is that traditional chess guidelines fail over and over again. One rule-of-thumb—"a queen against two knights is a draw"—turns out to be wrong most of the time. The queen actually defeats two knights in 89.7 percent of all possible positions. And, contrary to conventional wisdom, a queen defeats two bishops in 92.1 percent of possible positions (see Figure 6.17), and two bishops defeat a lone knight 91.8 percent of the time. But some of the winning lines of attack take more than 50 moves to accomplish, which brings into play a chess rule that a game is automatically a draw if 50 moves pass without a capture, a pawn move, or a mate.

The team behind Deep Thought didn't expect their machine to beat Kasparov. At best, they thought the machine might do well enough to force at least one game into a draw. The match's real importance was the opportunity for researchers to learn from a chess master. Most computer-chess researchers believe the world chess champion will eventually lose to a computer. The question is when.

Figure 6.17 By keeping its two bishops and king together, black was thought to form a group that the opposing king and queen could not penetrate. However, a computer analysis shows that the queen-king combination can win in almost every case, no matter where the bishops and king start. In the position shown, they win after 71 moves.

The answer may come with an IBM project aimed at creating an advanced chess machine—a successor to Deep Thought capable of evaluating a billion chess positions per second. Increasing the machine's speed is the most obvious and technically easiest way to make progress. Adding chess knowledge is more difficult.

The chess skills usually built into a program are generally very simple concepts that any amateur player would know, but those concepts combined with speed often create computers that play chess better than their human programmers. And programmers are trying to incorporate increasingly sophisticated chess knowledge. For example, most computer programs don't recognize the difference between a good and a bad bishop. In the game of chess, a good bishop is one whose movements are not impeded by its own pawns. A bad bishop is one that is restricted in the moves it can make because its own pawns are on the same color square it is, thus blocking it. A program that recognizes this situation can either avoid or correct it. By adding a simple check to determine the potential for this situation, problems can be avoided.

Hans Berliner, former world correspondence chess champion, has been working on one promising idea: chunking. Instead of looking at moves by individual pieces, Berliner's program treats logically related groups of pieces as units (see Figure 6.18). CHUNKER, as the program is called, then reasons about positions

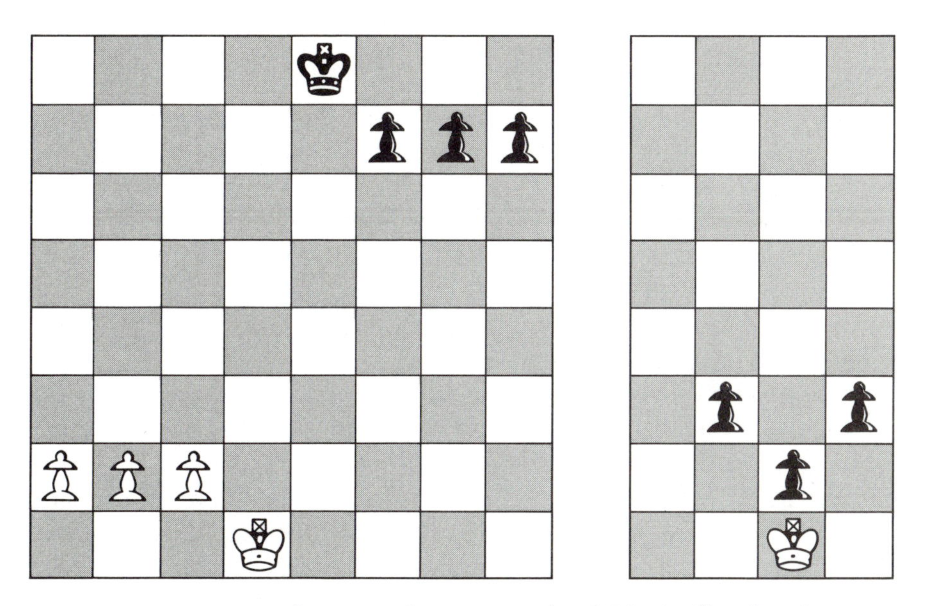

Figure 6.18 Human players are known to "chunk" logically related groups of chess pieces as units. The chess program CHUNKER applies similar logic to a game in which two players each have a king and three connected, passed pawns (*left*). The most important chunk type for analyzing this type of position consists of three back pawns and the white king (*right*).

that involve these units as they come up during a search. It relies extensively on a large library that lists each chunk type and its properties. As a result, CHUNKER can perform certain kinds of searches very quickly. An analysis that may take weeks of computer time to perform with a standard method takes only minutes with CHUNKER.

To Berliner, the chunking concept comes closer to the way people think about and play chess than the brute-search methods generally used in computer chess programs. So far, CHUNKER has been applied to a thorough study of a chess endgame in which each player is left with a king and three pawns. It has found errors in positions that have been studied for more than 300 years by human masters, and it regularly beats its authors from both sides of positions, playing either black or white. Within its limited domain, CHUNKER is an expert, playing the positions with a speed and accuracy that no human or machine has come close to matching.

Researchers are also looking into the possibility of designing a system that can learn from games played by expert human players. The usual difficulty is that expert players are very good at finding appropriate strategies and moves instinctively, and they often can't describe exactly how they do it, or express clearly how important one strategy is relative to another. One way to overcome that problem is to feed into the computer a large number of games between superior human players, letting the chess program figure out for itself which strategies are good and how much weight to give them.

Deep Thought already has a rudimentary kind of learning. By replaying completed games backwards, the machine has the capability of learning from its mistakes. Although it has trouble generalizing its new knowledge to situations that are similar but not identical to chess positions already encountered, the computer earns the respect of its opponents, who sometimes find its play surprisingly creative. It's the kind of behavior that leads some researchers to speculate whether there are other human endeavors in which creativity could be simulated by a clever, fast search.

But the gap between the way computers and humans play remains wide. Studies show that chess experts look at only a handful of moves and evaluate deeply just a few of them. They tend to rely on an instantaneous perception of a chess position as a whole. And the human mind's remarkable agility enables it to respond to expected situations. Computers don't have this kind of global view.

"I don't think any computer will be able to beat the world's best player before the end of this century," Kasparov says. The world's premier chess mind is confident that human ingenuity will come up with the strategies necessary to beat any chess computer, always keeping one step ahead, perhaps by learning from the way increasingly sophisticated computers play. "Chess is much wider than calculation. It's even wider than logic," Kasparov insists. "You have to use fantasy and intuition."

7

Shadows of Chaos

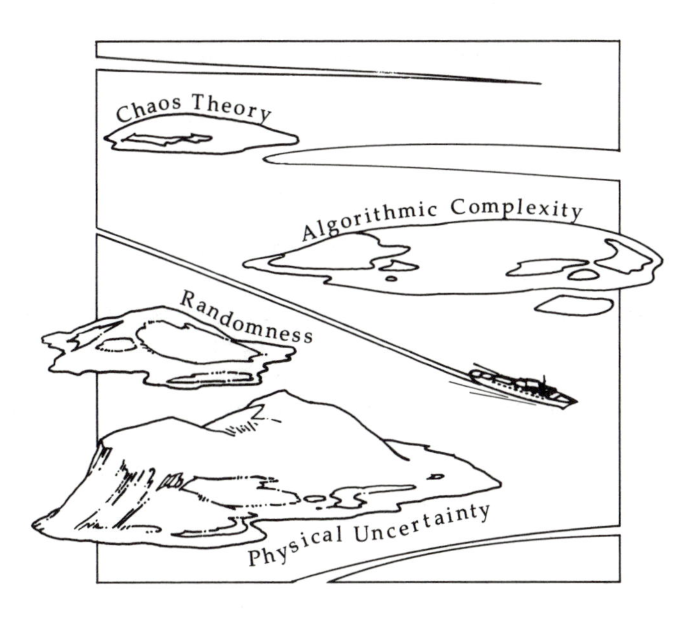

Calculating the result of a collision between two billiard balls is relatively easy. But when the number of particles is large, as in an unruly cloud of atoms or molecules, the system's behavior is so complicated that scientists must assume particle collisions are random. Therein lies a deep paradox at the heart of classical physics. On the one hand, the laws of physics are deterministic. They specify exactly what the outcome of any interaction should be. On the other hand, we are surrounded by physical processes that appear to be random.

The coexistence of randomness and deterministic physical laws poses perplexing riddles for scientists and mathematicians. How is it possible for a physical process to comply with both the deterministic laws of physics and the laws of chance? What is the source of randomness in physical systems?

Imagine a billiard table studded with large cylindrical bumpers in a regular pattern. When a ball ricochets through this configuration, each bounce magnifies any uncertainties in the ball's initial position and speed. The errors propagate so rapidly that no one could ever predict the ball's position after the first few bounces, on the basis of the table's geometry and the ball's initial velocity (see Figure 7.1). Although physical laws govern the ball's behavior, its movements rapidly become unpredictable. Small errors in knowledge of the ball's initial position and velocity grow exponentially with time, making long-term prediction of the future impossible.

With this kind of unpredictability evident even in a simple physical system, distinguishing between what is random and what is determined appears somewhat arbitrary. The spin of a roulette wheel, the throw of a die, and the toss of a coin are universally presumed to be random, even though physical laws govern exactly what happens. Weather, noise in electronic devices, and fluid turbulence are generally regarded to be the consequences of physical laws, despite their apparent unpredictability.

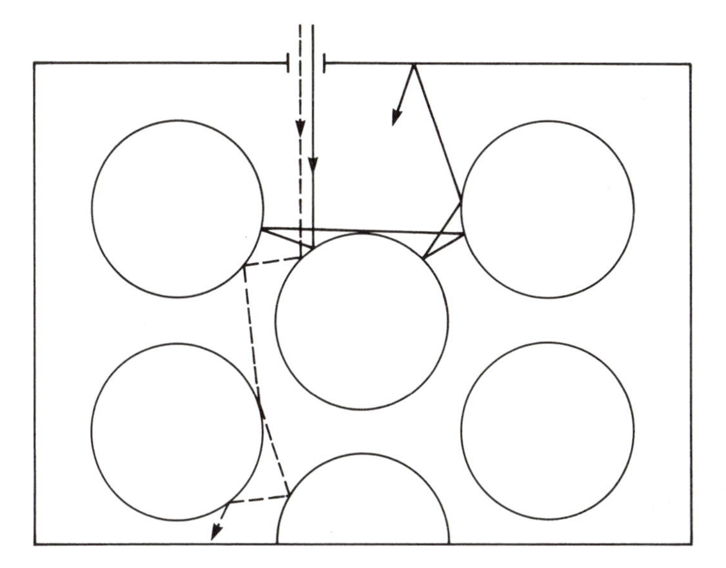

Figure 7.1 Balls with slightly different starting points end up following very different paths when they ricochet through a closely spaced array of cylindrical bumpers. This behavior demonstrates that any uncertainty in a ball's initial position makes it difficult to predict where the ball would be after a number of bounces.

Untangling the random from the determined is fraught with difficulties. Even a simple physical system can show incredibly complicated, unpredictable behavior. At the same time, randomness itself is an elusive concept. Books on probability rarely even attempt to define it. Questions of how to use randomness, how to create it, and how to recognize the real thing touch many areas in mathematics, science, and philosophy. They come up in lotteries and public-opinion polls, in the spread of randomness through a repeatedly shuffled deck of cards, in mathematical simulations of physical processes, and in the generation of fractal patterns.

Mixing Magic

Flipping a coin in the air, catching it, then determining whether it has come up heads or tails is to many people a prototypical random process. But coin tossing isn't really random at all. A mechanical gadget can flip a properly positioned coin so that the coin always lands showing the same face. Some magicians can make a coin come up heads on every toss.

Both gadget and magician count on the fact that a tossed coin obeys Newton's laws of motion. Each flip depends on the impulse given by the thumb to the coin and the distance traveled by the coin. If you could know the impulse given by the thumb in a particular case or had a well-defined mechanical flipper, then you could predict how the coin would fall. Any randomness would be not in the flipping itself but in how precisely the starting conditions are known.

In the physics of coin tossing, the most important parameters are the coin's upward velocity and its rate of spin. When the spin rate is low, the coin acts like a thrown pizza. It's unlikely to turn over, even if its upward velocity is very large and it travels a long distance. A coin may also come down without flipping over if it doesn't go high enough—even when it's spinning very rapidly. There would be too little time for the coin to turn over.

By calculating how often a coin turns over for a certain spin and upward velocity, one can predict whether it will come up heads or tails. The outcomes for a range of spins and velocities can be plotted on a graph (see Figure 7.2). Such a graph reveals that for the spins and velocities typically encountered in coin tosses, tiny changes in initial conditions make the difference between heads and tails.

Thus, coin tossing is almost random. A look at the spread in the way real people flip real coins indicates that heads and tails would each come up about half the time. But a slight bias would begin to show up after millions of tosses. The proportion of, say, heads would settle at a number such as .503 or .497, instead of getting closer to .500. Nevertheless, because the outcome is so

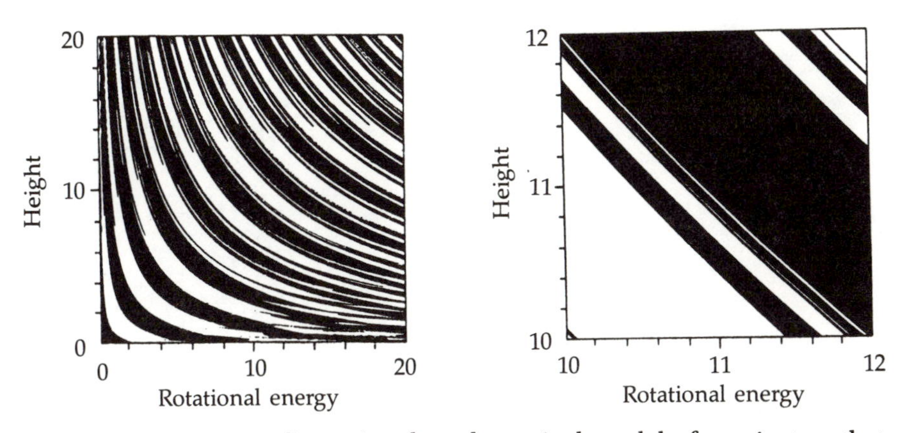

Figure 7.2 In a two-dimensional mathematical model of a coin toss that allows the coin to bounce from a hard surface before coming to rest, as the height (vertical axis) and spin (plotted as rotational energy on the horizontal axis) increase, the pattern of outcomes (black for heads, white for tails) grows more complicated (*left*). Magnifying a portion of the plot shows how small changes in the height and spin of a coin toss can change the result (*right*). A sequence of coin tosses will be random if the uncertainty in the initial conditions is large, compared with the width of the stripes.

sensitive to initial conditions, a vigorously tossed coin is random enough for most practical purposes.

Spinning a coin on its edge on a table is a different matter. Real coins tend to be lopsided, and which way a coin falls depends on the location of the coin's center of gravity. For certain U.S. coins, the bias is strong enough that heads come up 80 percent of the time. Such a bias is important when a tossed coin is allowed to land on a floor—a procedure many people wrongly think is more fair than catching the coin after a toss. Because a coin often spins around on the floor before it settles down, the results are more likely to be biased than if the coin were caught.

A close, hard look shows that many allegedly random processes aren't really as random as people suppose. A careful analysis of roulette reveals patterns that can be used to win money at a casino. Tumbling dice have characteristic movements that depend on the resiliency of the material from which the dice are

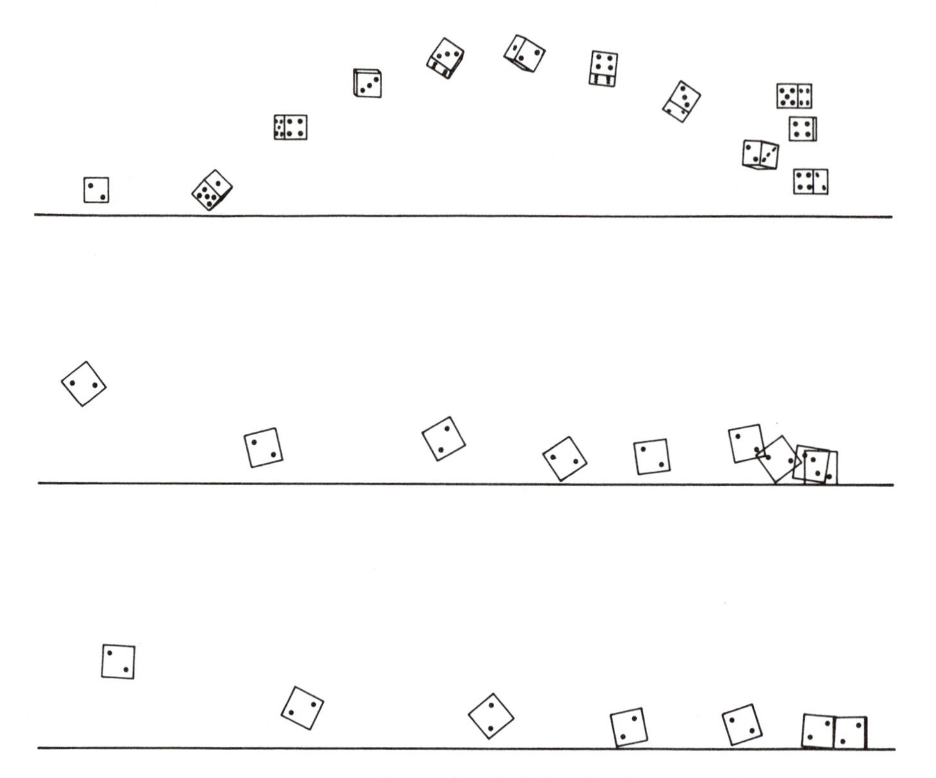

Figure 7.3 Casino operators detect loaded dice by dropping them carefully into a glass of water. A loaded die will tend to turn while descending through the water, whereas a fair die will sink with little rotation. These computer simulations show a fair die tumbling on all six sides (*top*), a loaded die constrained to tumble on just four sides (*middle*), and a fair die constrained to tumble on four sides (*bottom*). The motion of a loaded die tends to be slightly more erratic than that of a fair die.

made (see Figure 7.3). Even air-blown balls used to select numbers in a lottery aren't entirely unbiased. Indeed, the general problem of mixing, stirring, or shuffling things to ensure randomness is more complicated than most experts assume.

When the federal government carried out the 1970 military draft lottery, officials wanted to make the drawing appear as fair as possible. Scraps of paper carrying dates, one for each of the 365 days of the year, were stuffed into capsules, and all the capsules were placed in an urn, which was then shaken for several hours. However, when the capsules were chosen one by one from the urn, the dates were not random. Dates in December came up more often near the beginning of the list than dates from November, and so on. It turned out that capsules had been put into the urn according to a definite pattern. January dates went in first and tended to stay near the bottom. Because December dates went in last, they more often ended up near the top. Several hours of mixing were not enough to produce a random state.

Such problems have focused attention on the question of how long to stir up a mixture to ensure that its components are well mixed. It's a serious concern for manufacturers who need to know how long to mix the ingredients of products like pills and cakes. They want to avoid overmixing, both to save time and to avert the undesirable physical or chemical consequences of excessive agitation. Similar questions arise in the effort to calculate how long it would take a toxic gas to diffuse through a turbulent atmosphere.

Persi Diaconis, a statistician and professional magician, has taken a close look at the spread of randomness through a repeatedly shuffled deck of cards. He begins with a simple question: How many times must you riffle shuffle a deck of cards to ensure that the cards are randomly arranged? The answer isn't obvious, and tricksters have long taken advantage of this fact.

An old card trick, which depends on the perpetrator and victim being in different locations, illustrates the situation. To an unsuspecting victim, the perpetrator sends a new, ordered deck of playing cards and a set of precise instructions. The instructions tell the victim to cut the deck and shuffle the cards three times, using a riffle shuffle each time. Then the victim takes the top card, notes its identity, returns the card somewhere to the middle of the stack, and sends back the deck. Without any difficulty, the trickster finds what was once the top card—usually to the victim's astonishment.

It seems like an amazing trick, but it depends on the fact that three riffle shuffles are too few to mix a previously ordered deck of cards into an arrangement that no longer shows any patterns. Cutting and shuffling an ordered deck of cards one time leaves the deck with two interleaved chains. Each chain has cards in the same relative order as they started. Three shuffles produce eight such chains. In this particular trick, all one needs do to find the designated card is to deal out the cards as in a game of solitaire, putting them in chains as they come up. Eight piles will form, with one leftover card that doesn't fit any of the sequences.

How many times must a deck of cards be shuffled until it is close to random? According to Diaconis, it takes at least seven ordinary riffle shuffles before any trace of a pattern disappears. With five or fewer shuffles, the original order of the deck is strongly in evidence. With more than seven shuffles, the original order has virtually disappeared. Although no finite number of shuffles will ever make anything completely random, it doesn't take long to get close enough for practical purposes.

To come up with his estimate, Diaconis pioneered new mathematical methods of analysis that, unlike conventional techniques, provide useful answers to real-life problems. Conventional approaches simply prove that eventually things get well mixed. Such techniques establish how close to random a deck will be after, say, a million shuffles. But a card player wants to know the minimum number of shuffles needed for the original order to disappear.

The new techniques, which apply to a variety of shuffling processes, involve a mix of group theory (the mathematics of symmetry) and probability theory (the mathematics of disorder). The combination of these theories leads to remarkably precise estimates of how much mixing is sufficient to get a reasonable degree of randomness. It's somewhat ironic that mathematics invented to describe the orderliness inherent in symmetry can now be used to analyze its opposite—randomness.

The mathematical representation for card shuffling that Diaconis used doesn't duplicate exactly how an expert card handler shuffles a deck. But Diaconis found support for his theoretical result in Las Vegas, where, by law, dealers are required to perform five riffle shuffles and two other types of shuffles before a deck is ready for play. It would require a lot of work to take advantage of any patterns left after the shuffling process.

But human beings are often lazy. Players generally don't shuffle a deck of cards seven times. They're content with three, four, or five shuffles. This laziness (or ignorance) is something that experts can use to their advantage.

In the early 1970s, tournament bridge players started using computer-shuffled rather than hand-shuffled decks. Almost immediately, there was an outcry protesting the apparently wild fluctuations in the distribution of cards of different suits in computer-dealt hands. But the problem lay not in the computer but with human expectations. Subsequent research showed that hands in which suits were evenly distributed, that is, hands with four cards of one suit and three of each of the other suits, were actually more common in games in which people did the shuffling than they should have been according to theory. The reason for such a pattern is that cards during bridge play tend to clump together in groups of four of the same suit. Sloppy shuffling doesn't do enough to break up the groups, and when the cards are dealt out, the four players get a fairly even distribution of suits.

In fact, the intuition of bridge players had been shaped by generations of improperly shuffled cards. A close look at books published in the past reveals that many experts had figured out the problem, that cards generally aren't shuffled well, and had developed strategies that took such quirks into account. When computer shuffling was introduced, many of their strategies had to be revised.

One particularly striking feature of riffle shuffles turns out to be characteristic of many different mixing processes. Initially, shuffling seems to have relatively little effect on the arrangement of cards. Cards shuffled five or fewer times are not randomly mixed. Patterns are clearly discernible. Then after a few more shuffles, the patterns disappear quite abruptly, and the cards are mixed well. Further shuffling increases the degree of randomness only marginally. This property is somewhat unexpected because it's easy to imagine the original order gradually fading away (see Figure 7.4).

Almost all examples of mixing processes studied so far show an abrupt transition, but no one has yet come up with a theory to

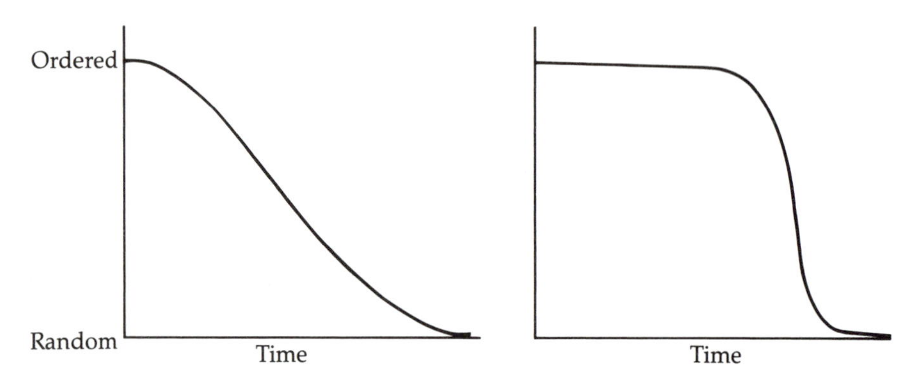

Figure 7.4 When shuffling cards, one would expect the deck to become more random gradually (*left*). However, mathematical analyses of shuffling show that the transition from order to randomness is quite abrupt (*right*).

explain why such behavior should occur. You might ponder the mystery in your own kitchen when stirring together white flour and cinnamon. At first, you see thick streaks as the ingredients mingle. After a few more strokes, the whole mixture suddenly smooths to a tan color.

Taking a Chance

Sequences of truly patternless, unpredictable digits are a valuable commodity. They're used by scientists and engineers for tackling a wide variety of problems, from developing secret codes for protecting sensitive information to modeling the behavior of atoms and complex systems. But everyone generally has to settle for numbers that fall short of true randomness.

The usual procedures for generating randomness—flipping a coin, rolling dice, shaking an urn, shuffling cards—are capable of producing random outcomes if they're done with appropriate precautions. But such methods are too cumbersome for everyday use by modern researchers, who prefer to use computers for quickly and efficiently generating vast stocks of random numbers.

The idea of using a computer—a predictable, logic-driven electronic machine—to create random numbers seems somewhat paradoxical. In fact, the role of computers in generating random numbers raises some unsettling questions.

Almost every type of computer has a built-in random-number generator. But truly random numbers are extremely hard to create. Instead, computers generate "pseudorandom" numbers according to fixed recipes. Most recipes even allow a computer to generate the same set of pseudorandom numbers over and over again. Such algorithms usually produce numbers that pass simple tests of randomness and mimic well the expected behavior of true random sequences. For example, on the average, a high number is followed by a lower one as often as a low number follows a higher one.

But, as more sophisticated tests of randomness show, no string of numbers generated by a simple computer process can be truly random. In fact, every scheme now in use for generating random numbers by computer has some kind of flaw. Often, the flaws are difficult to detect, and sensitive techniques must be used to discern the subtle patterns that may be hiding in vast arrays of digits.

Computer scientist George Marsaglia, a leader in devising increasingly powerful tests for randomness, has developed an array of tests for probing the randomness of number sequences. All currently accepted and widely used random-number generators fail Marsaglia's tests, meaning that the sequences they generate are not completely patternless.

In one test of a commercial random-number generator, researchers plotted the supposedly patternless numbers as points in a cube. When the cube was rotated slightly, the points suddenly appeared to line up into nearly perfect, parallel lines. Rotating the cube a little more improved the alignment further, revealing a nonrandomness that could not have been perceived by looking at the numbers themselves (see Figure 7.5). For some applications, such a pattern would make this particular random-number generator worthless.

The intrinsic lack of randomness in computer-generated sequences is troubling, especially because computer-generated ran-

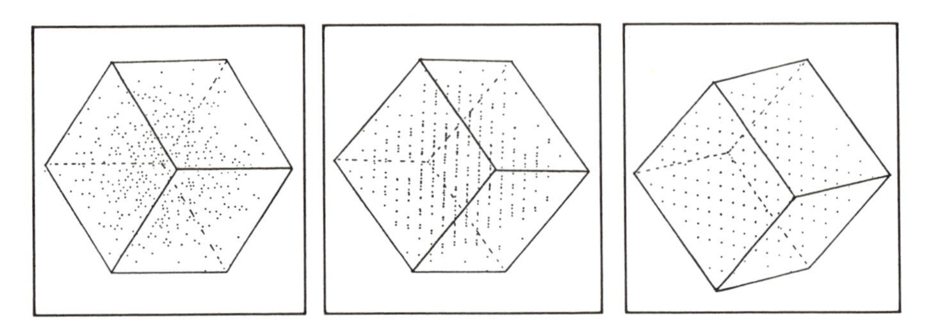

Figure 7.5 A particular sequence of computer-generated numbers seems random when plotted as points in a transparent cube (*left*). When the cube is rotated slightly, the points suddenly seem to line up in nearly straight, parallel lines (*middle*). Turning the cube more puts the points in an ordered pattern (*right*).

dom numbers are used extensively. Few researchers bother to probe the intricacies of the random-number generator they happen to be using. Most employ standard recipes that seem to have worked well in the past. In recent years, a small group of computer scientists, statisticians, mathematicians, and others, concerned that too many researchers use random-number generators blindly, has focused on identifying flaws in known random-number generators and inventing new, more reliable methods for generating random numbers.

Testing for randomness is difficult partly because it's the nature of randomness that anything can happen. Ironically, if a random-number generator always passes its tests, then it's probably flawed. To be truly random it would have to fail once in a while.

One alternative to using computers for generating random numbers is to use noise from a physical system, such as the electrical static heard between radio stations or the noise emitted by a geiger counter monitoring the radioactive decay of atoms. But it takes complicated equipment to sample and convert physical noise into digits. Moreover, the digits aren't reproducible—as they must be for many users of random numbers—and there's no proof that the source of physical noise itself is truly random.

Another, novel alternative for generating random numbers has recently emerged from computational complexity theory (see Chapter 6). The technique is based on the idea that certain mathematical operations are easy to perform but hard to reverse. One example is the multiplication of two large prime numbers. Multiplying the two numbers together is easy. But reversing the process to find the original two numbers by factoring is hard. Although researchers have developed a variety of techniques for factoring large numbers, they know of no method that guarantees any given number can be factored quickly.

The new technique involves starting with any number, multiplying it by itself, dividing the result by the product of two primes, then using the remainder to begin the process all over again, and so on, to generate a sequence of digits. The idea behind this scheme is a curious theoretical result linking the difficulty of factoring large numbers with the randomness of the scheme's outcome. In fact, researchers have proved that if factoring is computationally hard, the resulting set of numbers is indistinguishable from a truly random sequence. Conversely, if researchers can find some pattern or structure in the resulting sequence, they will in effect have a proof that there exist efficient schemes for factoring large numbers quickly.

Marsaglia and others are now subjecting the technique to a wide range of statistical tests to see whether it performs as well as theories predict. If it passes, the technique will be quite easy to put into use in computer software and in custom computer chips.

Strange Vibrations

Gyrating like a stiff but daring gymnast, the Space Ball moves in mysterious ways. This rudimentary toy's erratic oscillations show it to be one of a number of simple physical systems that reveal signs of chaos in their movements. Although such systems can be described by straightforward mathematical equations, their movements are quite unpredictable.

In essence, the Space Ball—a kind of double pendulum—is an efficient electric motor. A 9-volt battery hidden in the toy's base activates an electromagnet, which in turn "kicks" another magnet in the lower of a pair of suspended balls (see Figure 7.6). This clever combination of magnets and balls attached to stiff, swinging rods produces a wild mixture of unpredictable movements.

To study the toy's movements in detail, engineer Alan Wolf replaced the battery with a power supply capable of feeding in anywhere from 0 to 40 volts. By changing the voltage and the toy's starting position, he could sample the Space Ball's extensive, endlessly fascinating repertoire. He found that for some voltages and starting points, either nothing happens or the balls swing for only a brief period before coming to rest. For other voltages and starting points, the toy oscillates like a simple pendulum. But quite often, it executes complicated, erratic swings that seem to have no regular pattern. Sometimes it pauses for a short period of time, then suddenly starts moving again, spinning rapidly, only to slow

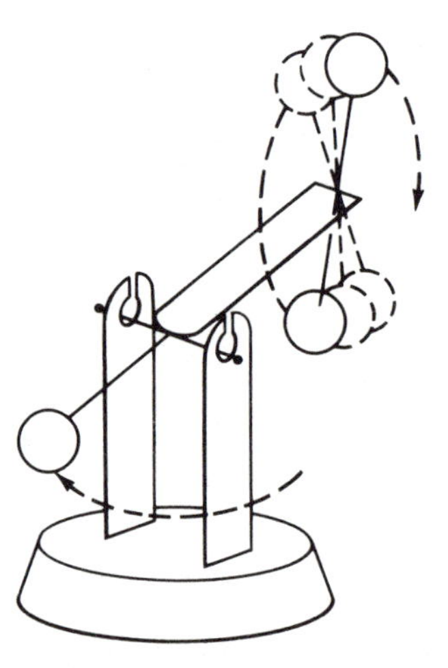

Figure 7.6 The Space Ball is an electrically driven toy that can exhibit the erratic oscillations characteristic of chaos.

down abruptly before resuming its gyrations. Says Wolf, "You can watch it for a week, and there may be no two periods of five minutes during which it does anything roughly similar."

Simple physical models like the Space Ball and the equations used to describe them help engineers and scientists to explore the new regime of chaotic dynamics and to test criteria needed to characterize it. Wolf has come up with two equations, based on Newton's laws of motion, that describe the toy's behavior. Solving the equations on a computer gives numbers that match the Space Ball's observed motions. Wolf's aim, however, is to quantify how much chaos is present in the toy, and also in any other system that may be suspected of exhibiting chaotic behavior. He has developed a computer program that calculates a quantity called the Lyapunov exponent. This number provides an estimate of how long the behavior of a system will be predictable. For a nonchaotic system, that exponent would be infinite because its future behavior is completely predictable. In chaotic systems, a tiny difference in starting conditions leads to widely divergent and, as a result, unpredictable behavior. The Lyapunov exponent puts a number on how fast this divergence occurs.

Another simple but useful physical model for a chaotic system is a mechanical device consisting of a small, vibrating table (constructed from a loudspeaker) and a ball that's constrained to bounce vertically on the table's surface (see Figure 7.7). The bouncing-ball system exhibits a veritable zoo of chaotic phenomena also seen in far more complex and consequently less comprehensible systems. Yet, as in the case of the Space Ball, a simple set of equations describes the physical system.

In the bouncing-ball apparatus, changing the table's frequency and amplitude alters the ball's motion. At certain frequencies, the ball's motion becomes extremely erratic. This model allows researchers to study how a physical system shifts into chaos. Moreover, because the ball makes a click every time it hits the table, listeners can actually hear the sound of chaos.

The bouncing-ball system also has educational value. Some people still attribute what is often labeled as chaos to factors such as background "noise" instead of believing that chaos results from the nature of the motion itself. Showing these skeptics a

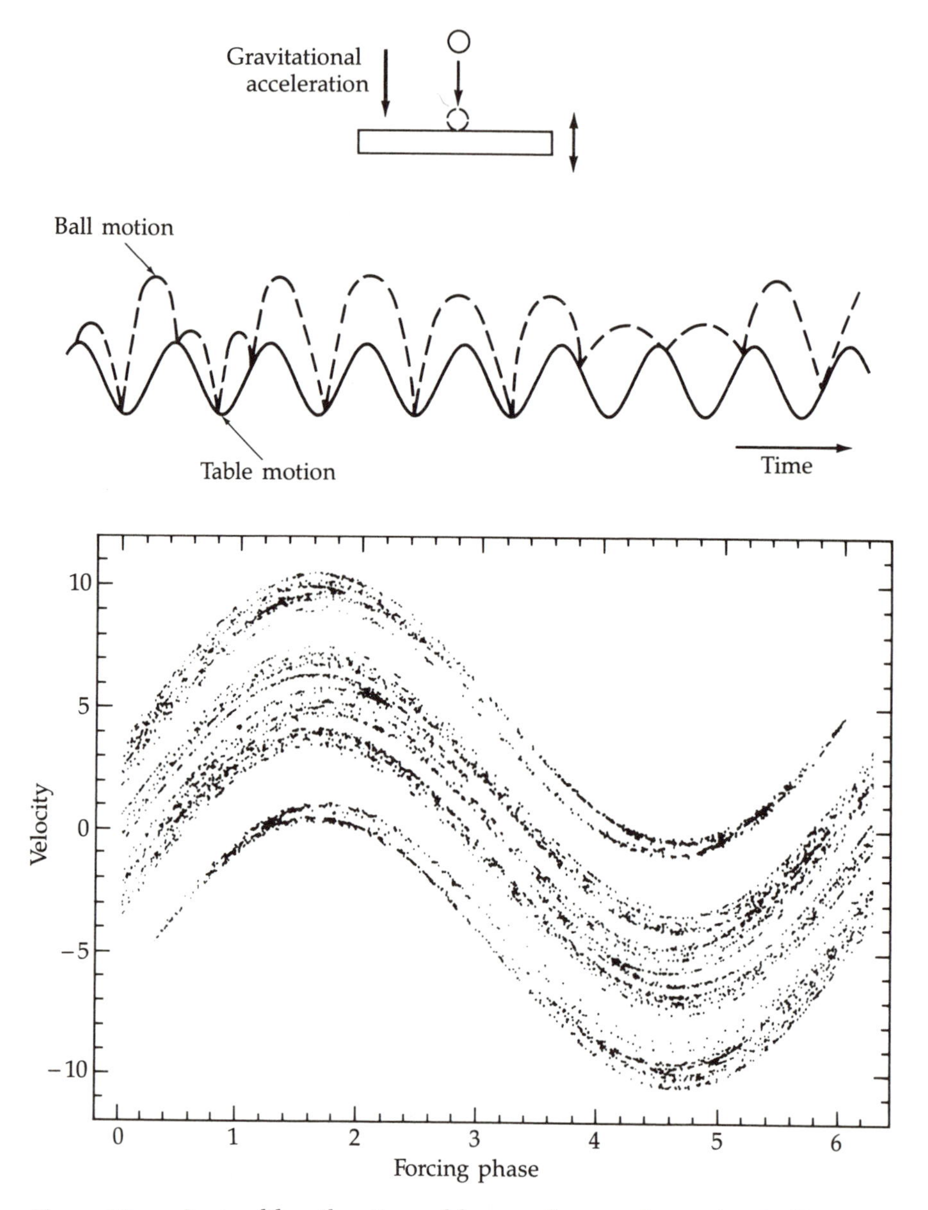

Figure 7.7 A steadily vibrating table can force a bouncing ball into a nonperiodic motion (*top and middle*). Each point in the plot (called a Poincaré map) represents the ball's impact velocity and time at a specific time (*bottom*). A Poincaré map, in general, is a collection of "snapshots" showing the system's state at discrete times and can be used to monitor chaos.

simple system that actually works as predicted mathematically can be very convincing.

Further insights into chaos come from detailed mathematical studies of the spherical pendulum, an ordinary, rigid pendulum mounted so it can move in any direction rather than just in a plane (see Figure 7.8). Although a pendulum is often considered the epitome of dynamical regularity, this one can behave chaotically if it's driven by applying a periodic force at its point of suspension.

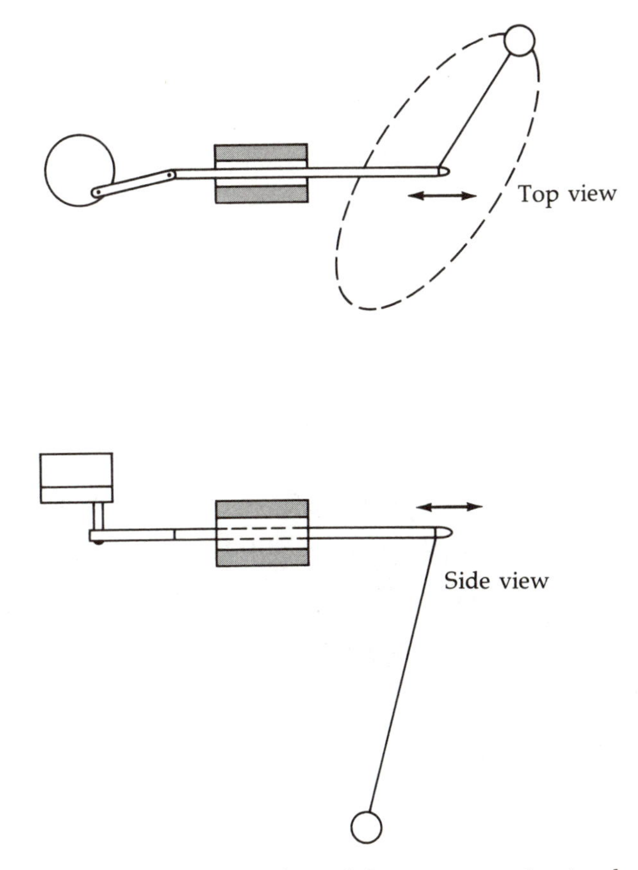

Figure 7.8 The forced spherical pendulum can consist simply of a ball on the end of a string. The ball is suspended so that it can swing in any direction. A motor forces the point of suspension to oscillate in a horizontal direction.

Although the pendulum is forced back and forth in one particular direction, its movements trace out a path on a two-dimensional spherical surface.

The spherical pendulum's bob executes either orderly or highly irregular movements, according to the frequency of the driving force. For example, with an applied driving frequency 1.015 times the pendulum's natural frequency, the pendulum, starting from rest, begins to oscillate back and forth with a larger and larger amplitude. Within 30 seconds, the motion becomes two-dimensional and ultimately settles down into a stable pattern. Lowering the driving frequency to 0.985 times the pendulum's natural frequency changes its resultant motion drastically. The pendulum's movements follow no discernible pattern (see Figure 7.9). At any given instant, the pendulum may be found in linear, elliptical, or circular motion over a wide range of amplitudes.

Even when it's not chaotic, a driven spherical pendulum displays interesting effects. In the long run, its bob settles into any of

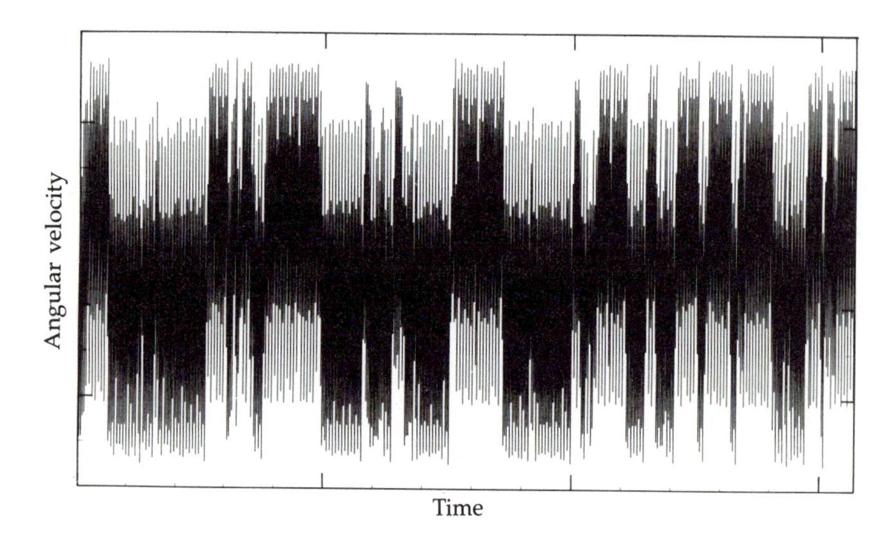

Figure 7.9 A forced, damped pendulum can exhibit chaotic motion, as seen in this erratic plot of angular velocity versus time.

four different patterns of regular behavior, comprising various types of clockwise and counterclockwise periodic motions. By computing solutions of differential equations describing the pendulum bob's motions, researchers can create a kind of map specifying into which pattern the pendulum ends up for a range of starting positions and velocities.

The resulting picture, in which different colors represent the four different possible types of motion, calls to mind an incredibly mixed up plate of spaghetti (see Color Plate 13). It shows a tremendous amount of structure that doesn't disappear when the picture is magnified. You see swirls within swirls within swirls on all scales. For a wide range of initial conditions, the pattern of colors is so complicated that predicting into which pattern the pendulum will settle is practically impossible. Often, you're so close to a boundary between two colors that it's extremely difficult to identify which color regime you're in.

Such complexity in a physical system as simple and well characterized as the spherical pendulum is a startling revelation. Clearly, the existence of complex, intricate behavior doesn't require complicated fundamental principles. Even very simple systems, described by simple equations, can generate behavior showing an extraordinary degree of complexity and diversity.

Stalking the Wild Trajectory

Simple physical systems such as the spherical pendulum can exhibit chaotic behavior. A snapshot of the system reveals its state at one particular instant. What happens next is reasonably predictable, but the system's long-term behavior is not. The reason for this loss of predictability is the way uncertainties escalate. Researchers associate such chaotic processes with mathematical procedures in which small errors made at successive steps rapidly accumulate to destroy any semblance of a pattern.

Consider the following simple mathematical procedure. Start with a number less than 1, then keep doubling it. Every time the answer is greater than 1, lop off the 1 and retain only the decimal or fractional part of the number. For example, if the starting number is $1/3$, the sequence goes: $1/3$, $2/3$, $4/3$ (which, according to the rules, would be rewritten as $1/3$), $2/3$, and so on. The sequence has a definite repeating pattern.

Now suppose that the computer or calculator can't handle fractions and expresses all such numbers as decimals rounded off to a specified number of places. Thus, the fraction $1/3$ might be expressed as .33. The sequence becomes: .33, .66, 1.32 (which is rewritten as .32), .64, .28, .56, .12, .24, .48, .96, and so on. It no longer has a simple repeating pattern. Increasing the precision to three decimal places produces the sequence: .333, .666, .332, .664, .328, .656, .312, .624, .248, and so on. By the ninth step, the sequence has again diverged significantly. Increasing the precision to a larger number of decimal places doesn't help. The errors still accumulate rapidly and predictability disappears.

The rapid growth of errors in repeated mathematical operations makes the numerical study of mathematically modeled physical systems exceedingly difficult. The problem involves untangling the consequences of rounding error in mathematical calculations, uncertainties caused by limitations in measurements of physical systems, and the intrinsic characteristics of the physical system itself. It shows up with a vengeance when mathematical functions are repeatedly evaluated, or iterated. On the surface, iteration appears quite straightforward, even mindless: calculate the value of a given mathematical expression, or function, for some initial value; then substitute that answer back into the original expression to get a new value, and so on. This simple iterative process often leads to surprisingly complex, unpredictable mathematical behavior.

Chaotic behavior turns up in the iteration of many different functions. For certain functions, successive points can be plotted on a graph to produce a two-dimensional cloud of dots. The sequence of plotted points—one dot leading to the next dot—corresponds to an orbit, or trajectory. Researchers term such an orbit chaotic if it jumps erratically from dot to dot, never settling down into any kind of regular pattern, even though the motion tends to stay within a bounded region and some neighborhoods may be visited more often than others. A tiny shift in starting point produces a very different sequence of dots, although the overall dot pattern remains roughly the same.

But computation is intrinsically inexact. If a small change in starting point leads to rapidly diverging results, then errors made when rounding off numbers during a computation may also influence the results. How much does the appearance of chaotic behavior depend on calculator or computer limitations?

Consider a computer working with numbers to an accuracy of 14 decimal places. Computer experiments show that two neighboring orbits starting at points differing only in the last decimal place will look totally unrelated after a few dozen steps (see Figure 7.10). For some iterated functions, it's not unusual for the distance between orbits to double at every step.

These results imply that a tiny error in rounding off at the first step is sufficient to destroy any attempt at predicting where the orbit is likely to be after, say, 50 iterates. On top of that, errors in rounding occur not just at the beginning but at every step.

The question of predictability is significant because the iteration of appropriate mathematical expressions is a standard method for finding approximate solutions of equations used to describe the dynamical behavior of materials, fluids, and other physical systems. For example, scientists have developed sets of equations for modeling atmospheric processes to predict changes in weather patterns and climate. Often, their predictions are based on the results generated by iterating equations thousands of times. How good can those predictions be if the initial conditions are generally known only to one or two decimal places and the answers coming out of a computer may be intrinsically uncertain? Similar problems arise when researchers compute the way a metal may fracture or how air sweeps past an airplane wing.

On the positive side, researchers already have reasons to believe that chaotic orbits are more than just numerical artifacts. Different computers calculating to different numbers of decimal

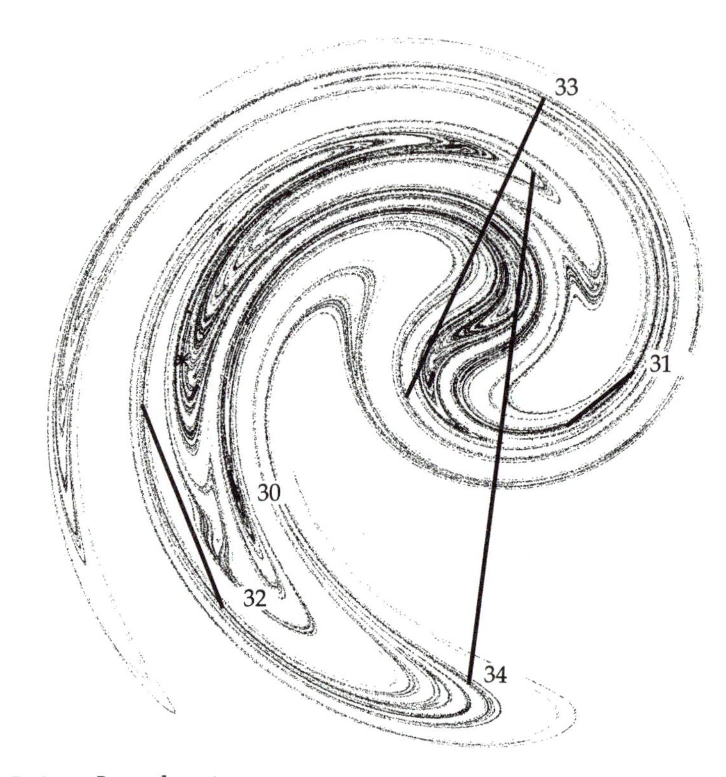

Figure 7.10 In a chaotic process, two neighboring orbits starting at points differing only in the fourteenth decimal place (marked by the asterisk) will be widely separated (shown by the lines) after as few as 30 steps. The divergence of nearby orbits is the underlying reason chaos leads to unpredictability.

places produce similar pictures. Although the details vary, the overall features stay the same.

Mathematician Jim Yorke and his colleagues have found a way to separate computer error from intrinsic unpredictability. Like detectives closely shadowing their errant quarry, they track the erratic hops of chaotic processes, carefully checking to see how closely each step in such a process sticks to a "true" path. Their rigorous mathematical procedures prove there exists a true orbit that stays near the computed, "noisy" orbit of a given chaotic process for a long time.

The idea is that while a numerical orbit will diverge rapidly from the true orbit (one calculated exactly without any error) with the same initial point, there often exists a true orbit with a slightly different initial point that stays near, or shadows, the computed (noisy) orbit dot by dot for a long time—for as many as 10 million steps if computational errors are no larger than the fourteenth decimal place.

The procedure represents a rigorous determination of how long a true trajectory stays near a numerical one. That's done by keeping close tabs on round-off errors. The computer does all the necessary arithmetic. As it calculates a trajectory, the computer places a carefully constructed numerical box, within which a true orbit must lie, around each point. When it proceeds to the next point in the trajectory, it carries the box in a somewhat distorted form along with it. If the original box and the new box overlap in just the right way, then at least one true trajectory stays boxed near the numerical trajectory (see Figure 7.11).

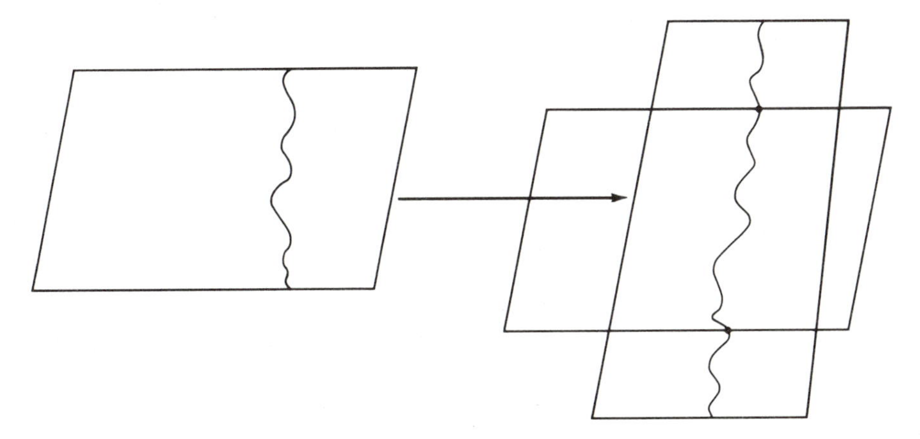

Figure 7.11 As it calculates an orbit, the computer constructs a small numerical box about each point (*left*). When it jumps to the next point, it carries a distorted form of the box with it (*right*). Verifying that a noisy orbit stays close to a true orbit means checking that the new box properly overlaps the old box at each step.

If the computer has sufficient numerical precision, the boxing scheme can be extended to at least the first 10 million steps. Such long shadowing times are striking when compared to the high rate at which orbits diverge from each other. However, at some point, the computer encounters a "glitch" at which successive boxes don't overlap, meaning that the true and computed trajectories start to diverge significantly. The errors suddenly refuse to stay neatly boxed. From that point on, there's no way to certify that the computed orbit remains close to a true one.

Yorke and his colleagues proceed on a case-by-case basis. There is no a priori guarantee that their procedure will work for arbitrary initial conditions. However, for any specific trajectory —for a given function and starting point—the results can be checked for a certain number of iterates with Yorke's method.

For example, Yorke has demonstrated that a true trajectory passes through every one of the millions of iterates producing the array of dots in a figure known as the Ikeda map (see Color Plate 14). That figure represents the results of iterating an equation describing the electromagnetic field within a ring-shaped laser cavity. Yorke is also interested in the statistical behavior of the Ikeda map, which shows that trajectories spend more time in some regions (shown as brighter areas) than in others. The trajectory may stay in the bright regions for several thousand iterates, then suddenly escape to the dim halo region for 10 or 20 dots before being pulled back into the light. One mathematical objective is to try to describe these random escapes.

Although Yorke's work does certify that for 10 million or more points, specific chaotic orbits are real rather than merely numerical artifacts, many questions remain. For example, what is the ultimate behavior of chaotic trajectories when there are infinitely many points? For a given function, do different trajectories always form roughly the same pattern of dots?

Recent work on chaotic systems demonstrates that chaos represents predictability in the short run coupled with unpredictability in the long run. Small errors in knowledge grow exponentially with time, making long-term prediction of the future impossible. But there are many questions yet to be answered. Chaos theory is a new, vigorously expanding field. It's possible the most important questions have yet to be asked.

Troubling Uncertainty

Where does the randomness necessary for the dynamical behavior of molecules in a gas come from if the universe is at heart a deterministic system? Chaos theory offers a possible answer to this question, a bridge between the random world of diffusing gases and roulette wheels and the mechanical, clockwork universe governed by Newton's laws of motion.

We now know that deterministic systems, which can be expressed by precise mathematical equations, can behave in such a complicated way that no one can predict exactly what they will do in the future. The best anyone can do is to make probabilistic statements about them. What's striking is how many physical systems appear to fall into this category, suggesting real limits on what we, as humans, can learn about our physical environment.

Consider a system as simple as the sun, the Earth, and one other, smaller body, such as an asteroid. Although all three bodies gravitationally act on each other according to Newton's laws of motion, the complex influences of the two larger bodies can make the asteroid's movements so erratic that its future position is describable only in terms of probabilities (see Figure 7.12). The problem gets even more complicated with a greater number of bodies, and researchers are looking for evidence of chaos in the orbits of satellites around planets such as Jupiter and in the orbit of the planet Pluto itself.

Finding chaos doesn't mean it's impossible to make predictions. Space scientists can pinpoint and predict planetary locations and velocities well enough to plan space missions years in advance. They could feel confident that Neptune would be right where it was supposed to be when the Voyager planetary probe finally arrived after years of travel. In the same way, astronomers

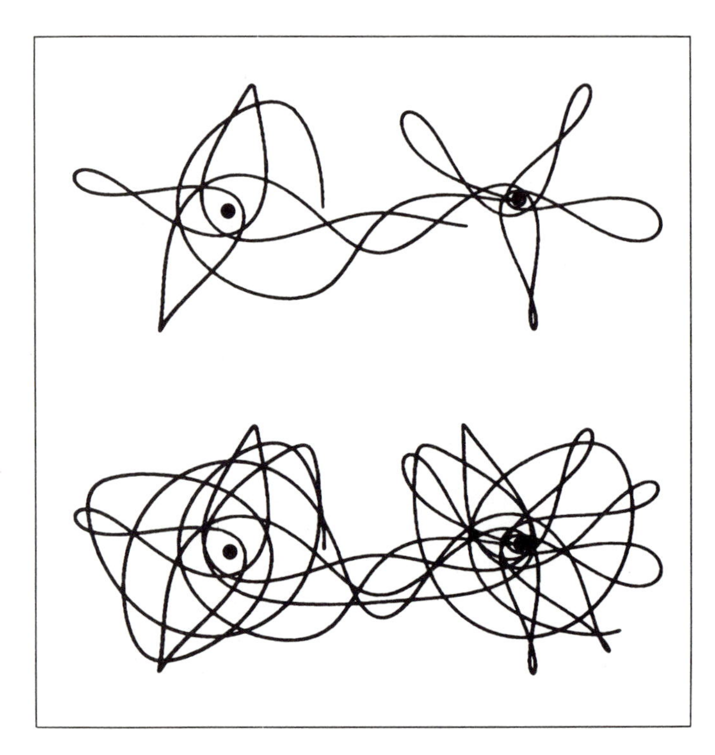

Figure 7.12 This computer-generated path for a small planet orbiting two suns shows how complicated such an orbit, which involves three gravitationally interacting bodies, can get. The upper plot shows the orbit's first stages; the lower plot continues the motion for a longer period of time.

can predict solar and lunar eclipses centuries in advance of their occurrence.

Even then, locations and velocities are known only to a certain degree of precision. In solving equations of motion, small errors accumulate, and predictions become less reliable as the time gets longer. Where Pluto will be a billion years from now is anyone's guess. Chaos theory gives researchers an appreciation of how little complexity in a system is needed to produce complicated phenomena.

The idea that simple dynamical systems can lead to complicated types of behavior isn't entirely new. A century ago, French mathematician Henri Poincaré caught a glimpse of this complexity when he tangled with solutions of the equations of motion for three interacting bodies—the notorious three-body problem. In 1908, he pondered the connection between chance and determinism:

> A very small cause which escapes our notice determines a considerable effect that we cannot fail to see, and then we say that the effect is due to chance. If we knew exactly the laws of nature and the situation of the universe at the initial moment, we could predict exactly the situation of that same universe at a succeeding moment. But even if it were the case that the natural laws no longer had any secret for us, we could still only know the initial situation *approximately*. If that enabled us to predict the succeeding situation with the *same approximation*, that is all we require, and we should say that the phenomenon had been predicted, that [it] is governed by the laws. But it is not always so; it may happen that small differences in the initial conditions produce very great ones in the final phenomena. A small error in the former will produce an enormous error in the latter. Prediction becomes impossible, and we have the fortuitous phenomenon.

Poincaré may have suspected or even understood how much of nature is chaotic, but only recently has this awareness seeped into the general consciousness of the scientific community. Many scientists are just beginning to realize that a clockwork universe really has very little to do with the real world. Chaos theory suggests a very different picture, in which almost all dynamical systems can, under the right conditions, exhibit chaotic behavior. It has already had an impact on a broad spectrum of scientific disciplines, including ecology, economics, physics, chemistry, engineering, and fluid mechanics. Researchers have detected signs of chaos in the swirling movements of heated fluids, the yearly swings in insect populations, the turbulence responsible for Jupiter's Great Red Spot (see Figure 7.13), and the exotic behavior of oscillating chemical reactions. Chaotic effects may limit the extent of reliable weather forecasting.

In chaotic systems, uncertainties grow at an escalating rate, sharply restricting how far into the future one can make a precise prediction about the system's behavior. Thus, the unpredictability of chaotic motion is not a matter of ignorance. Gathering more

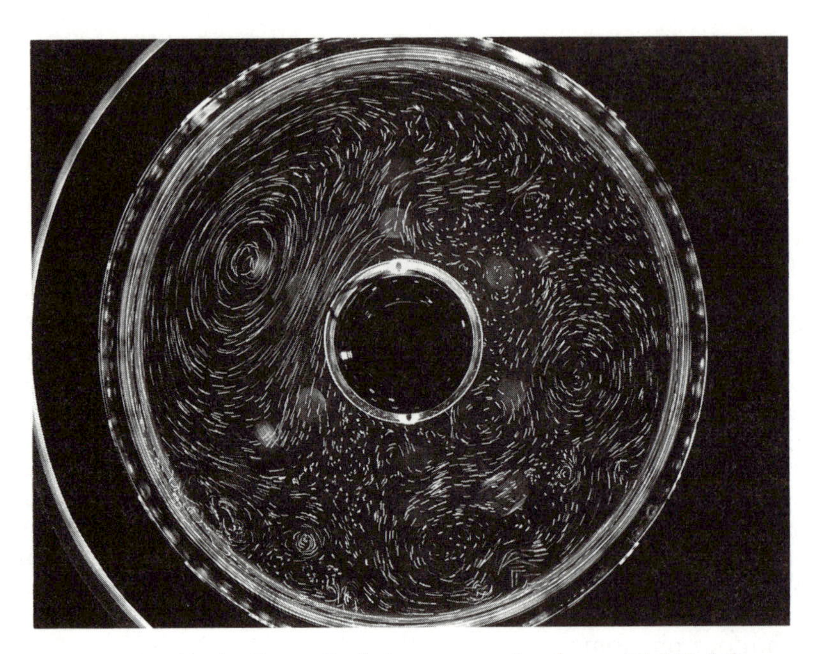

Figure 7.13 Jupiter's Great Red Spot extends about 25,000 kilometers, roughly one-third of the radius of the planet. Smaller spots circle it every 7 days. In a laboratory experiment to simulate the formation of such large vortexes, researchers pump fluid into a rapidly rotating, cylindrical tank. At a sufficiently high pumping rate, a jet of water begins to flow in a direction opposite to that normally expected in a rotating system. Above a certain pumping rate, the jet is unstable and breaks up into one or more vortexes. When more than one vortex is initially present and the current is strong enough, the vortexes merge. Here, plastic beads suspended in water outline a large, stable vortex with many of the characteristics of the Great Red Spot.

information or making more precise measurements doesn't eliminate the system's long-term unpredictability. No computer could ever keep pace with the rapidity at which such a system itself evolves. The only way to see what happens is to observe the system itself. There is no shortcut.

Because strictly deterministic systems can to varying degrees mimic true randomness, chaos offers scientists a way of understanding complicated behavior as something that is purposeful and structured instead of extrinsic and accidental. In systems that normally behave according to well-understood and well-accepted physical rules, researchers have found instances in which the solutions to equations describing such systems unexpectedly show chaotic characteristics. Looking more closely, they discover the real world does indeed behave as the solutions predict, even when those solutions are chaotic. This gives researchers hope that they eventually may develop detailed analyses of some physical systems that previously seemed open only to probabilistic solutions.

Some people who have studied and thought deeply about chaos insist there is more to come. Physicist Joseph Ford believes chaos theory will fundamentally change our view of the world "by forcing us to face our limitations." Scientists typically assume that science is closed, consistent, and single-valued. In other words, ask a question of nature, and nature provides only one answer, which is part of some complete theory. Yet the mathematical logic that forms a foundation for science is neither closed nor consistent. Mathematician Kurt Gödel proved that any mathematical system is incomplete—that there will always be questions that can be asked but not answered for any particular logic system.

Chaos theory suggests there are unanswerable physical questions. In other words, nature is so complicated that some questions, even apparently simple questions, have answers so complex that to specify them would require more information than that contained in all human logical systems combined. Ironically, such hard-to-answer questions are often easy to ask.

Computer scientist Gregory Chaitin has explored how randomness is connected with Gödel's ideas, number theory, and knowledge about physical systems. He is one of several researchers who have contributed to a new theory, known as *algorithmic complexity theory*, aimed at providing measures of the fundamental complexity of systems by looking at their information content.

One of the deepest questions in all probability theory is how to establish that a given string of digits, such as the decimal representation of π or numbers generated by a computer, is truly random. To get a handle on this question, one can look at the strings of numbers used to label points on the number line. The interval from 0 to 1 contains an infinite number of points, each specified by a number drawn from an infinite collection of decimal numbers. Some of these numbers are fractions, which have either finite decimal expansions or infinitely long decimal expansions in which a group of digits repeats periodically. But some decimals cannot be expressed as fractions. Irrational numbers, such as π, have decimal expansions that follow a never-ending, erratic sequence.

Similarly, in binary notation, all numbers between 0 and 1 can be expressed as infinite strings of ones and zeros. Conversely, every conceivable string of ones and zeros, in whatever combination, corresponds to a point somewhere in that interval. Some of those strings are easy to express. For example, .1111111111. . . can just as easily be expressed by the statement "Keep on writing down ones" or by a simple computer program telling the computer how to print out the digits. Similarly, .01010101. . . has a simple, easy-to-express pattern. However, most strings of digits can be expressed only by writing down the entire string (see Figure 7.14). There's no formula or short way of describing them. The simplest way to specify such sequences is to provide copies of them. The digits in these strings are said to be random. According to this definition, a random sequence is one for which there is no compact description.

But what is the connection between this and earlier definitions of randomness? Because randomness implies a certain lack of order, and because disorder can occur in infinite variety, no single test, expressed as an algorithm of finite length, can rigorously prove a sequence to be random. For real numbers expressed as decimals, this notion of randomness means that almost all real numbers are random and can be expressed in binary form as a random sequence of ones and zeros. But it can't be proved that a particular sequence is random. A random sequence has too much information; each successive bit provides fresh information, and the amount of information becomes endless.

To relate this concept to the physical world, imagine reducing the description of any physical system to a sequence of digits. For

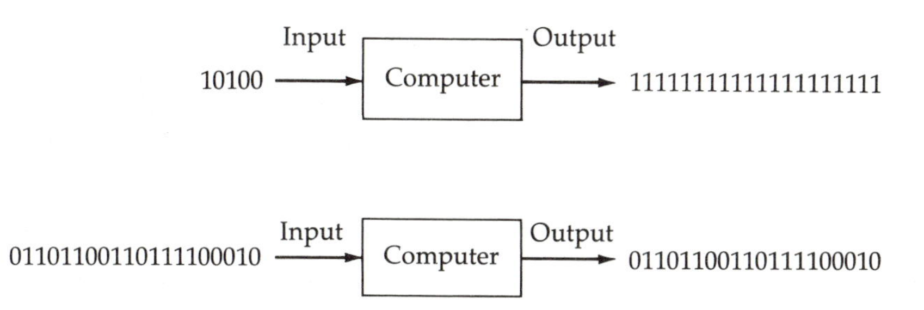

Figure 7.14 The algorithmic definition of randomness relies on the capabilities and limitations of the digital computer. In order to produce a particular output, such as a series of binary digits, the computer must be given a set of explicit instructions—an algorithm. If the desired output is highly ordered, as in the top configuration, a relatively small algorithm will suffice. A series of 20 ones, for example, might be generated by some hypothetical computer from the program 10100, which is the binary notation for the decimal number 20. For a random series of digits (*bottom*) the most concise program possible consists of the series itself.

example, you could express the positions and velocities of all molecules in a given physical system as a string of ones and zeros. The idea is to regard a physical system's initial conditions as input information for a computer simulation of its future behavior. In ordinary systems, a small amount of information is converted into a large quantity of output information in the form of reasonably accurate predictions. For chaotic systems, such a simulation is pointless. The computer is reduced to being merely a copying machine. At best, the system generates as much information as the amount put in. The computer can describe the system in increasingly greater detail, but it can't make any worthwhile predictions (see Figure 7.15).

The implications of this direct connection between chaos and randomness are startling. Gödel specified a human limitation on what can be known and proved in mathematics. Chaos theory suggests limits on how much humans can learn about physical systems. The answers to some questions are so complicated that scientists will never be able to find them, no matter how big their computers.

Observations	Predictions	Theory	Size of theory
0101010101	01010101010101010101	Ten repetitions of 01	23 characters
	01010101010000000000	Five repetitions of 01 followed by ten 0's	44 characters

Figure 7.15 In a mathematical analysis of reasoning as employed in science, a scientist's observations can be represented as a sequence of binary digits. Such observations are to be explained and new ones to be predicted by theories, which are regarded as algorithms instructing a computer to reproduce the observations. The task of the scientist is to search for algorithms that are as simple as possible. In the example, the better theory explaining the observations is the shorter one. If the data are random, the minimal programs are no more concise than the observations and no theory can be formulated.

Classical physics was based on the conviction that the future is determined by the present, and therefore a careful study of the present permits an unveiling of the future. Although this unlimited predictability was never more than a theoretical possibility, in some sense, it was an essential element of the scientific picture of the physical world. The real irony is that nonchaotic systems, which form the basis of our present understanding of nature, are actually scarce.

For centuries, scientists deemed randomness a useful but subservient citizen in a deterministic universe. The situation is greatly changed today. Algorithmic complexity theory and nonlinear dynamics, which includes the study of chaotic systems, together establish the fact that determinism actually reigns over a finite domain. Outside this small haven of order lies a vast, largely uncharted wilderness of randomness, in which determinism barely shows its face.

In the domain of chaos, specifying initial conditions requires superhuman skills, including the ability to compute infinite algorithms of maximum complexity, the ability to make infinitesimal distinctions, and absolute intolerance of the slightest error. From a practical point of view, determinism is at best only a temporarily local property that vanishes under an avalanche of escalating error. In principle, nature can be both deterministic and random. In practice, determinism is a myth.

To some extent, physicists have learned to live with similar fundamental limitations. For example, Werner Heisenberg specified that in quantum mechanics, one can't simultaneously measure the position and the momentum of a particle to an arbitrary precision. If one quantity is known exactly, then the other cannot be. According to Joseph Ford, chaos theory goes further by suggesting that no variable—no measurable quantity—can be measured to an arbitrary precision. Thus, all systems have a level of "noise" that can never be eliminated.

A large number of provocative questions at the frontiers of chaos theory remain unanswered. The random element of chaotic systems endows scientists with the freedom to explore a vast range of physical behavior patterns. If physicists start playing with the limitations on human knowledge suggested by chaos theory, they could be driven toward new insights into the nature of our world, just as the developers of quantum mechanics were in the 1920s and 1930s. It means building intrinsic uncertainty into anything a scientist would want to measure.

8

Truth and Beauty

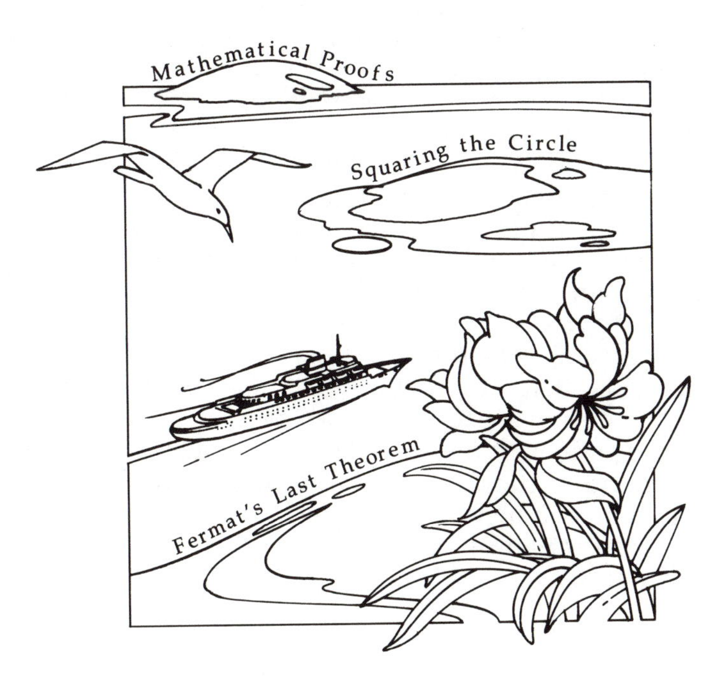

Research, whether scientific or mathematical, usually begins with the gathering of data that enable one to guess that certain statements may be true. The second step involves establishing the truth of those conjectures. Whereas experiment is the final judge in science, the bottom line for mathematicians is the logical consistency of their results, expressed in proofs of theorems.

The pursuit of mathematical truth is a tricky business, full of mysteries and surprises. Mathematicians prove results that don't seem to make sense. They construct some proofs so large and complicated that only computers can handle them. Like philosophers, they tangle with questions so simple that a 10-year-old can understand them, yet so difficult that no one has yet found answers.

Mathematicians have also learned to be wary. They know that logical systems can't provide all the answers. In the early 1930s, mathematician Kurt Gödel proved that any sound, consistent formal system incorporating arithmetic contains statements about mathematics that cannot be proved. Before this startling result appeared, mathematicians had expected that all of mathematics, when it was suitably formalized, would turn out to be complete.

As Gödel proved, mathematics is only as true as its foundations, and it's impossible to prove that its foundations are completely solid. It's the nature of mathematics to pose more problems than it can solve. Mathematicians may believe that certain statements are true without necessarily being able to follow a formal, step-by-step process to determine their truth.

Conceivably, mathematics may also contain true statements that are theoretically provable but the proofs required are too long and involved for a human being to cope with them. Indeed, mathematics may be built on little islands comprising the pieces of mathematics that can be validated by relatively short proofs, leaving an ocean of possibilities beyond the reach of human effort.

The Straight Side of Circles

There's no telling where thoughts about a seemingly simple, even trivial, question may lead. Consider the problem of turning a circle into a square. Cut a circle out of a sheet of paper. Then cut the circle into pieces so that the pieces, when fitted back together, form a square having the same area as the original circle. The task seems impossible: How do you get rid of the curves? But there is a mathematical solution. In 1989, Hungarian mathematician Miklös Laczkovich accomplished the mind-bending feat of proving that it is theoretically possible to cut a circle into a finite number of pieces that can be rearranged into a square.

The problem solved by Laczkovich is related to an ancient riddle known to Archimedes, Euclid, and other Greek scholars. At issue is whether it's possible, with just a ruler and compass, to draw a square with an area equal to that of a given circle. This particular problem remained unsolved for centuries despite the efforts of numerous mathematicians, both amateur and professional. As a matter of fact, general interest in the problem was so great and the number of proposed but erroneous proofs so voluminous that in 1775, the Paris Academy found it necessary to pass a resolution stating that no more purported solutions to the circle-squaring problem would be examined.

In the end, the solution hinged on the properties of the number π. A circle and a square have equal areas only if the ratio between a square's side and a circle's radius equals the square root of π. In 1882, mathematicians proved that π is what is known as a transcendental number, effectively ruling out the possibility of constructing a square out of a circle using only ruler and compass. There's no way to divide up a line in the required ratio because all ruler-and-compass constructions are the geometric equivalent of algebraic equations, and π, as a transcendental number, can't be the solution to such equations.

Laczkovich tackled a version of the problem originally devised in 1925 by mathematician and philosopher Alfred Tarski. Tarski removed the ruler-and-compass restriction and asked whether there is any way to cut up a circle into pieces that can be rearranged into a square of the same area. In the previous year, Tarski and Stefan Banach had proved a remarkable analog of the same conjecture in three dimensions, showing paradoxically that a sphere could be cut up and rearranged not only into a cube of the same volume but also into a cube of twice the volume. In fact, a sphere sliced up in just the right way could be rearranged into virtually any shape of any size.

At first glance, the Banach-Tarski result sounds like a contradiction. The explanation of the paradox lies in the nature of the pieces that get rearranged. The pieces aren't solid chunks with nice boundaries. Instead, they are so convoluted, diffuse, and intertwined that it's mathematically impossible to measure the volume of an individual piece. It's only when the pieces are put together that the resulting solid has a measurable volume. And, strangely, there's no contradiction in concluding that this can be done in different ways, resulting in different volumes.

The two-dimensional version of the problem turned out to be more difficult to handle than the three-dimensional version. Mathematicians who studied Tarski's circle problem strongly suspected there is no way to cut up a circle to make a square without losing even a single point out of the circle. Banach himself proved that any rearrangement of a figure in the plane must have the same area as the original. In 1963, Lester Dubins, Morris Hirsch, and Jack Karush proved the problem couldn't be solved by cutting a circle into "ordinary" pieces with well-behaved, relatively smooth boundaries—the kind you make using a pair of scissors—no matter how many such pieces are used.

Laczkovich proved that "squaring the circle" is possible, provided that the pieces have the right form. His pieces encompass an array of strange, practically unimaginable shapes. Although some resemble those in an ordinary jigsaw puzzle, others are curved segments, strangely twisted bits, or collections of single, isolated points. Remarkably, assembling a square from these pieces of a circle is possible simply by sliding the pieces together. No piece has to be rotated to fit into place. The resulting square has no gaps and no overlapping pieces. Laczkovich estimates that about 10^{50} pieces are needed, many more pieces than there are water molecules in all the world's oceans combined. Further analysis may bring that figure down considerably.

Laczkovich's proof applies not only to circles but also to nearly any plane figure with a mathematically well-behaved boundary. Any such figure can be cut into pieces and rearranged into a square of the same area with no gaps or overlays. For example, it's possible to take apart a triangle or an ellipse and fit the pieces back together as a square without even rotating any of the pieces.

The solution of the squaring-the-circle problem shows that such apparently simple notions as length, volume, point, curvature, and area must be treated with due respect and care. In particular, the new result bears on the fundamental questions of what mathematicians really mean by the notion of curvature and how they decide when two objects have the same area. The proof shows that present ideas about area seem to be correct. Cutting one shape into pieces, then rearranging the pieces into another shape, is a reasonable way to show that the shapes have equal areas. But curves and straight lines are so different that they can be converted into each other only with the use of strange manipulations.

Beyond Understanding

"Don't touch this problem. It's too difficult. You may not get anywhere, and you may never graduate."

That was the advice computer scientist Clement Lam received about 20 years ago when he was a graduate student. Lam heeded the warning but eventually came back to the problem, a well-known unanswered question in the area of mathematics known as *combinatorics*. In 1988, working with a team of computer scientists, Lam completed an elaborate, extensive computer search that apparently settled the question. At the same time, the effort raised important issues concerning the nature of mathematical proof.

Like many notoriously difficult mathematical problems, Lam's problem is relatively easy to state. Imagine a square table consisting of 111 columns and 111 rows. The aim is to fill 11 of the spaces in each row so that all columns also have exactly 11 spaces filled. In addition, for any pair of rows selected, both rows must each have only one filled space that falls in the same column. The second condition is what makes the problem particularly difficult.

A completed table, with filled spaces represented by ones and empty spaces by zeros, is one way of expressing what is known as a "finite projective plane." This particular example, if it exists, would have the designation "order 10." Such structures play a role in the design of switches for complicated communications networks so that data flowing in and out of a system are properly routed.

The search for finite projective planes of a particular order has a long history. Mathematicians know that such planes, or equivalent tables, are simple to construct when the order is a prime number, such as 2, 3, 5, and so on, or a power of a prime, such as 4, 9, or 27 (see Figure 8.1). Although it's been conjectured that these are the only possibilities, no one is certain. Consequently, mathematicians have been looking for exceptions, starting with order 10 (the smallest unresolved nonprime order).

Lam's computer search depended on a combination of careful analysis and a cleverly designed computer program. With more than 470 trillion ways of filling in just one of the 111 rows, trying every possible combination was out of the question. In 1970, F. J. MacWilliams, Neil J. A. Sloane, and John G. Thompson applied various mathematical techniques to narrow the search to a few, big cases. Lam and his team used those ideas as the basis for their attempt at solving the problem.

Even after narrowing the search considerably, the researchers still faced a gigantic computational task that seemed barely feasible on the fastest available computers. An early result added even more bad news. The researchers learned from their analysis that if a solution exists, it has no symmetries, which makes it much harder to find any solution that may exist.

Lam and his colleagues began the first leg of the computations in 1979, using a minicomputer to run through the simplest

	A	B	C	D	E	F	G
1	1	1	0	1	0	0	0
2	0	1	1	0	1	0	0
3	0	0	1	1	0	1	0
4	0	0	0	1	1	0	1
5	1	0	0	0	1	1	0
6	0	1	0	0	0	1	1
7	1	0	1	0	0	0	1

Figure 8.1 This square lattice consisting of seven rows and seven columns is an example of a finite projective plane of order 2. Each row and each column have exactly three points, represented by ones. In addition, any pair of rows or columns has only one point in common. There are five other possible arrangements that match these requirements.

cases. Those efforts required almost a year of computer time. Realizing that the final case would take more than 100 years on their minicomputer, Lam and his collaborators looked for help. Computer scientists at the Institute for Defense Analyses came to the rescue. They agreed to run Lam's program on a Cray supercomputer whenever the computer happened to be idle. In the end, the search consumed about 3,000 hours of the Cray computer's time, spread over 2 years. If the researchers had been required to pay commercial rates for the computer time, the bill would have come to millions of dollars.

In the end, Lam's exhaustive computer search established that there are no finite projective planes of order 10. In other words, no combination of ones and zeros satisfies the rules for such a plane. Lam's result for order 10 fits the established pattern.

Whether mathematicians now regard the order-10 question as settled depends on how comfortable they are with the notion of such a large, complex, computer-based proof. Some mathematicians are skeptical, and many await some form of confirmation. Lam himself regards his effort as more like an experimental result than a definitive proof. Successfully applying the techniques he developed to solve this problem to other problems would help vindicate the work.

Lam's results are not a proof that anyone can check easily. Even trying to verify them on another computer seems impractical. Several people had a hand in writing the original computer programs. The computations were done in bits and pieces over a number of years and required large amounts of computer time. Moreover, computers themselves are fallible. Cosmic rays or a flaw in a computer's operating system can easily change a calculation.

Even if everything worked the way it was supposed to work and the plan was carried out successfully, a proof that something doesn't exist is still far more difficult to verify than finding an example satisfying the given criteria. Was the search really exhaustive? Was every possible case checked?

Equally perplexing and disturbing is the apparent trend toward long, complicated proofs for settling even apparently simple questions in mathematics. Some mathematicians believe that computer-assisted mathematical proofs are bound to come up more often in the future. And a number of mathematicians are still not satisfied with the computer-aided proof of the four-color problem, performed more than a decade ago. The question is whether four colors are always enough to fill in every conceivable map that can be drawn on a flat piece of paper so that no countries sharing a common boundary are the same color (see Figure 8.2). The proof filled several hundred pages, and a computer played a vital role in verifying certain necessary facts. No individual or

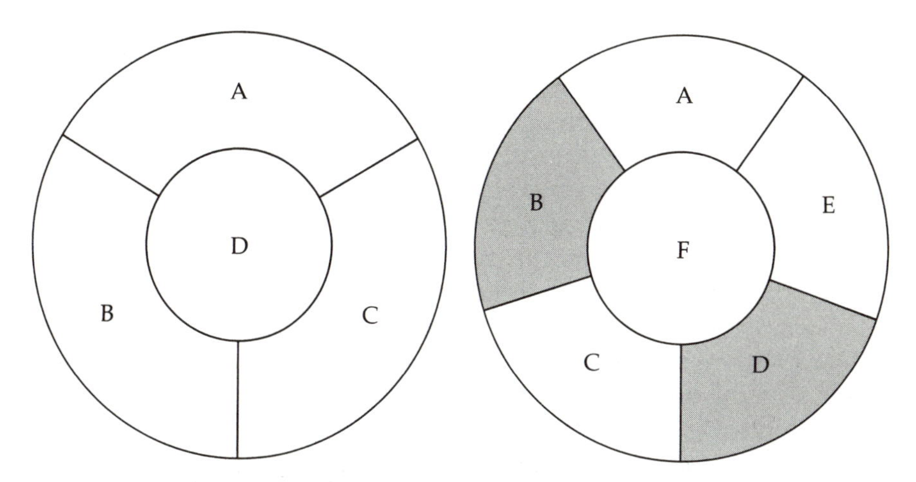

Figure 8.2 Three colors are not enough to color all maps, as is shown by the map of four countries in which each country is adjacent to the other three (*left*). On the other hand, it is not correct to assume that the number of colors needed to color a map is the same as the highest number of mutually adjacent countries in the map. In the map at the right, for example, no more than three countries are mutually adjacent, and yet four colors are needed: three for the outer ring of countries and one for the country in the center.

group of persons could check the entire proof without the aid of a computer.

What about a proof that runs to more than 10,000 pages, spread across 500 articles in mathematics journals? That was the result of one of the greatest mathematical quests of the last few decades: the classification of all finite simple groups. If you think of a group as a molecule, then the simple groups are its atoms. Simple groups are the building blocks of all groups that have a finite number of elements. They play much the same role in group theory (see Chapter 3) that prime numbers play in number theory and that chemical elements play in chemistry. Understanding simple groups is the key to understanding all groups.

But simple doesn't mean easy to find. Solving this enormous, highly complex puzzle required the efforts of more than 100 mathematicians all over the world. The project started in the nineteenth century with the classification of simple Lie groups, which are defined in terms of transformations of spaces having a specific number of dimensions. Mathematicians also discovered so-called exceptional groups, which conform to no obvious pattern yet are mysteriously related to a variety of mathematical operations. Exceptional groups have applications in mathematical physics, with connections to quantum mechanics, quantum field theory, and the nature of space-time.

The list of finite simple groups grew slowly during the early 1900s. Mathematicians added a few new members, but nothing really striking happened until 1957, when they discovered several new avenues worth exploring and developed a strategy for continuing the search (see Figure 8.3). By the 1960s, lengthy computer calculations were often needed to pin down the existence of the larger, more elusive members. But not even a computer could handle the largest "sporadic" group, the last type of finite simple groups to be investigated. That particular group, now called "the monster," was finally constructed in 1982 from a group of rotations in a space having 196,883 dimensions.

The completed list represents what is probably the largest group effort in the history of mathematics. But the effort involved more than simply establishing a complete list of objects satisfying some specified conditions. It was also an attempt to understand the intrinsic structure of a mathematical system. Along the way, mathematicians made important discoveries affecting a wide range of fields: mathematical logic, geometry, the theory of computer algorithms, and number theory.

Normally, proofs of significant, new theorems are carefully checked by experts and later read by a broader spectrum of mathematicians. Gradually, the theorem becomes part of the body of accepted mathematical knowledge. Yet no individual has read the proof of the classification theorem, probably the biggest theorem ever seen in mathematics, in its entirety. Indeed, some of the papers themselves were so long that journal editors had to assign separate portions to different persons for checking.

Even the subject of errors in proofs isn't entirely clear-cut. Mathematicians found several minor errors in the classification

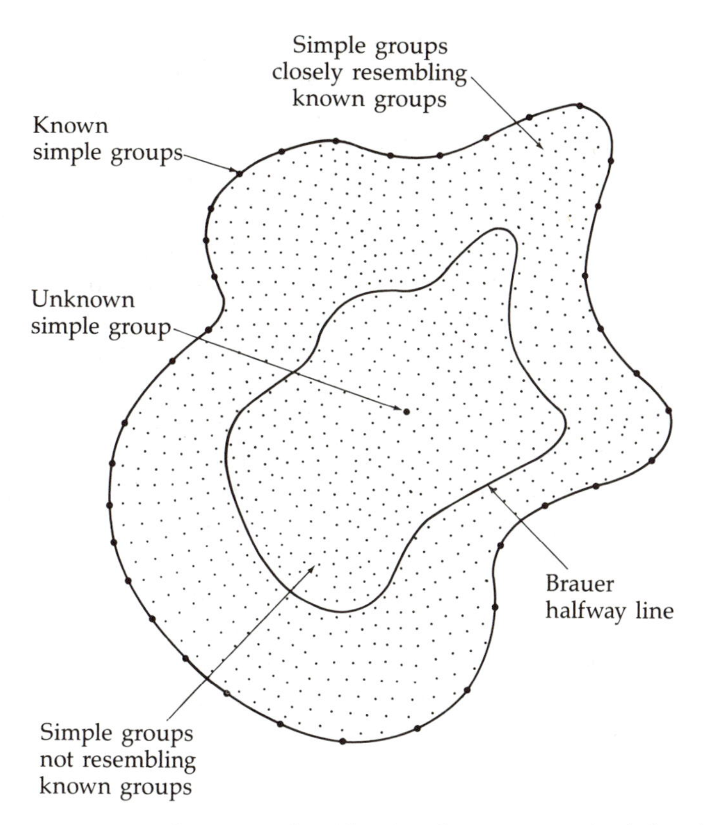

Figure 8.3 Proving the group classification theorem required showing that any simple group with an arbitrary, unknown subgroup structure (dot in center) can be linked with known simple groups (dots at the boundary of the enclosed region), thereby bringing it closer to the boundary. The work of Richard Brauer led to a criterion that describes the stage about halfway to the boundary. Mathematicians had to explore roughly 100 paths from the center to the boundary to complete the proof.

theorem, which were easily corrected. Such blemishes are not uncommon. In one recent case, a number of errors in the first few pages of a proof discouraged a serious check of the proof's validity. Yet, when the full proof was finally checked, the errors turned out to be minor and easily corrected, and the entire proof was valid. At the same time, most mathematicians can cite at least one case of an apparently proved theorem that actually turned out to be false. No one is sure that the classification theorem contains no errors whose consequences would be great enough to threaten the integrity and validity of the entire proof.

What makes the group classification theorem exceptional is that the proof's great length forces the checking process to continue long after the initial announcement of success. Mathematician Daniel Gorenstein describes the present, uncomfortable situation: "Most simple-group theorists are fully convinced that the 'essential' shape of the existing classification proof is correct. Perhaps the best evidence of its overall validity is the fact that, despite some earlier predictions to the contrary, no simple groups have been found since the announcement!"

Mathematicians are now working to put the classification theorem for finite simple groups on a sound footing. That means revising the original proof to construct a simplified "second-generation" proof. Gorenstein says, "The sounder the health in which we leave the classification theorem at this time, the more accessible we can make it, the greater will be the chances of fundamental new ideas or even alternate approaches being discovered in the future."

A Marginal Note

The most famous unsolved problem in mathematics started out as a note scrawled in the margin of a book (see Figure 8.4). Now known as Fermat's last theorem, the problem was first proposed

Figure 8.4 Fermat's last theorem was first set down in the margin of *Arithmetica*, a work on number theory by the ancient Greek mathematician Diophantus and translated into Latin by C. G. Bachet. Fermat had studied the book closely, making many marginal notes in his copy of it. After Fermat's death in 1665, his son published a new edition of Bachet's translation of *Arithmetica* that included Fermat's marginal notes in an appendix. The title page of the book is shown in this illustration.

DIOPHANTI ALEXANDRINI ARITHMETICORVM LIBRI SEX,

ET DE NVMERIS MVLTANGVLIS

LIBER VNVS.

CVM COMMENTARIIS C. G. BACHETI V. C.

& obſeruationibus D. P. de FERMAT *Senatoris Toloſani.*

Acceſſit Doctrinæ Analyticæ inuentum nouum, collectum ex varijs eiuſdem D. de FERMAT Epiſtolis.

TOLOSÆ,

Excudebat BERNARDVS BOSC, è Regione Collegij Societatis Ieſu.

M. DC. LXX.

by seventeenth-century French mathematician Pierre de Fermat. In a sentence that was to haunt mathematicians for centuries to come, Fermat noted that although he had a wonderful proof for the theorem, he didn't have enough room to write it out. After Fermat's death, scholars could find no trace of the proof in any of his papers, and ever since, mathematicians have struggled in vain to solve the problem.

Fermat's conjecture (as it should properly be called until a proof is found) is related to an observation by the Greek mathematician Diophantus that there are positive integers, x, y, and z, which satisfy the equation $x^2 + y^2 = z^2$. (For example, $3^2 + 4^2 = 5^2$.) In fact, this equation has an infinite number of positive-integer solutions (see Figure 8.5). Fermat proposed that there are no solutions to the equation $x^n + y^n = z^n$, when n is greater than 2.

A century after Fermat's death, Leonhard Euler proved that Fermat's conjecture is true for $n = 3$. Later mathematicians found proofs for other special cases, and a computer search performed a decade ago showed that Fermat's last theorem was true for all exponents less than 125,000. Therefore, if a counterexample

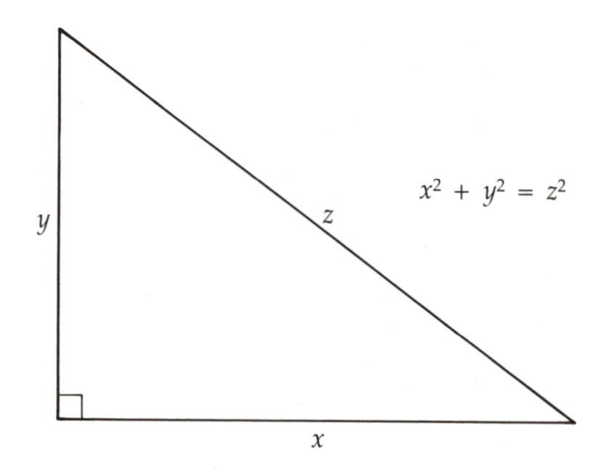

Figure 8.5 According to the Pythagorean theorem, the lengths of the sides in a right-angled triangle with sides x and y and hypotenuse z are related by the equation $x^2 + y^2 = z^2$. For triangles with sides of integer length, the smallest Pythagorean triple is 3, 4, and 5.

exists, it would involve huge numbers. But a proof for the general case has remained elusive.

One effort to tame Fermat's conjecture started with the work of Gerhard Frey, who happened to be looking at equations for elliptic curves, written generally in the form $y^2 = x^3 + ax^2 + bx + c$, where a, b, and c are constants. He found a way to express Fermat's last theorem as a conjecture about elliptic curves. That put Fermat's problem into an area of mathematics where a wide range of tools and techniques had already been developed for solving problems.

Frey wrote down the elliptic curve that would result if Fermat's conjecture is false. That curve turns out to have peculiar properties, and mathematicians examining the curve's equation had the feeling that the curve could not exist. If that could be proved, then Fermat's conjecture would have to be true. Frey started on this task, but he failed to fill all the gaps in his attempted proof.

Then French mathematician Jean-Pierre Serre suggested that one of his own conjectures, if proved, would help patch up Frey's effort. In 1987, Kenneth A. Ribet worked out the necessary proof of Serre's conjecture for a large class of situations. Ribet ended up showing that Fermat's theorem is true if certain elliptic curves arise from so-called cusp forms.

This approach ties Fermat's last theorem to a central question in number theory, known as the *structural conjecture*. It states that all elliptic curves arise from cusp forms. Most number theorists, for a variety of convincing reasons, believe the structural conjecture to be true, although it has not been proved. Even without a proof, the conjecture plays a major role in number theory. In fact, a proof that the conjecture is false would come as a shock to the mathematics community.

There may still be other possible approaches to proving Fermat's last theorem. In 1988, the centuries-long search for a proof suddenly seemed at an end when Japanese mathematician Yoichi Miyaoka proposed a proof for a key link in a chain of reasoning establishing the conjecture's truth. Miyaoka's method, built on work done by several Russian mathematicians, linked important ideas in three mathematical fields: number theory, algebra, and geometry. Though highly technical, his argument filled fewer than a dozen manuscript pages, which is short for such a significant, modern proof.

In 1983, Gerd Faltings showed that if there are any solutions to Fermat's equations, then there are at most a finite number of solutions for each value of n. Some of the key ideas for Faltings' proof came from the work of Russian mathematician S. Arakelov, who was looking for connections between prime numbers, curves, and geometrical surfaces. Both Arakelov and Faltings found that analogs of certain classical theorems already well established for geometrical surfaces could apply to curves and provide information about statements, such as Fermat's last theorem, that involve only integers.

Then A. N. Parshin, following Arakelov's lead, proved that if the arithmetical analog of an inequality, or bound, governing certain geometrical structures were true, then Fermat's last theorem would also be true. That inequality, in its original form, had been discovered by Miyaoka and Shing-Tung Yau. To solve the Fermat problem, Miyaoka, a specialist in algebraic geometry —which concerns the relationship between geometric surfaces and solutions of equations—ventured into a relatively new field known as arithmetic algebraic geometry. In this discipline, mathematicians look at surfaces that result when only integer solutions of equations are considered. Miyaoka tried to complete the chain of reasoning leading to a proof of Fermat's last theorem by showing that the so-called Miyaoka-Yau inequality, in a modified form, also fits an analogous case for equations with integer solutions.

But careful scrutiny of Miyaoka's complex proof turned up several flaws that cast doubt on the proof's validity. There was no straightforward way around the difficulties. Changing some statements in one part of the proof forced changes in many other parts, pulling apart the carefully constructed argument.

That Miyaoka's initial attempt failed is hardly surprising or unusual in mathematical research. Normally, mathematicians privately circulate proposed proofs and discuss possible errors or oversights for months before they gain enough confidence to announce a proof publicly. In Miyaoka's case, the fact that Fermat's last theorem was such a famous unsolved problem put him in a spotlight that he had not sought.

Although every attempt to prove Fermat's last theorem or to find a counterexample so far has failed, the efforts have led to important new mathematical techniques that could be applied to other problems. And gradually, Fermat's last theorem has become linked with other important questions, which has already made the pursuit of Fermat's last theorem of considerable value to mathematics.

Mathematicians are fairly confident that someone, if not Miyaoka, will eventually come up with a proof of Fermat's last theorem. But there's still the nagging worry that perhaps Fermat's last theorem may be one of those mathematical statements that either cannot be proved or has such a long proof that no human being will ever be able to find it.

Gregory Chaitin, who played a key role in developing algorithmic complexity theory, concludes that proving whether a certain class of equations has finitely or infinitely many solutions is often intractable. "Such questions escape the power of mathematical reasoning," he insists. "This is a region in which mathematical truth has no discernible structure or pattern and appears to be completely random. These questions are completely beyond the power of human reasoning. Mathematics cannot deal with them."

Math of the Spheres

To outsiders, mathematicians seem to conjure up their concepts out of nowhere. But most mathematicians insist that the mathematical structures they find are not artificial creations of the human mind but are as natural and real as those theoretical structures formulated by physicists to describe and "explain" the real world.

However, whereas physicists traditionally have searched for a single logical structure to account for natural phenomena, mathematicians explore the space of all possible logical structures, only a part of which overlaps with the real world. The flights of a mathematician's imagination, though unconstrained by physical experiment, are instead constrained by the need for consistency: the need for any new piece to fit with all other parts, including those that connect with the real world. Thus, if mathematics is about structures that, like the concepts of theoretical physics, are a part of the real world, then it isn't surprising that mathematics is also an effective tool for analyzing the real world. Some mathematical discoveries are bound to be useful is science.

Nevertheless, the issue of whether mathematics is discovered or invented has a long history and has never been settled to everyone's satisfaction. Since the time of Plato, the more popular epistemological thread has been to believe that mathematics exists independently of human knowledge. The idea is that, in some sense, mathematical structures already exist, and mathematicians are charged with finding the right tools to unveil them.

On the other hand, many philosophers, including Immanuel Kant, and a number of mathematicians have maintained that the ultimate truth in mathematics lies in the fact that its concepts are constructed by the human mind. This difference in viewpoint leads to different ideas about what is important in mathematics and how to go about doing mathematics.

The favored belief that mathematics is discovered rather than invented has led to the popular notion that it could serve as a universal language. If human beings ever encountered another intelligent form of life in the universe, the two civilizations would likely share a basic mathematics that might well serve as a means of communication.

But it's conceivable that the central core of our mathematics has little in common with anyone else's. Although human predilections and historical accidents wouldn't change the truth of theorems, they might dramatically influence the course taken by mathematical research, and how the results obtained are organized. In other words, our mathematics may be more arbitrary than most people would like to think.

Even the use of mathematics as a tool for understanding the laws of nature is not straightforward. Although the laws of physics are not set by humans, neither does nature proclaim them unambiguously. Conceivably, there may be more than one way to formulate nature's laws in a mathematical form. At the same time, the intrusion of physics into mathematics strongly influences the historical development of mathematics.

Furthermore, not all of mathematics is strictly the result of logical necessity. The peculiar nature of the human mind also enters the picture. For instance, mathematicians often link logical thinking with idiosyncratic preferences for short proofs and formulations, which they term "elegant" or "beautiful."

One way of picturing the structure of mathematics is to imagine the steel skeleton of a massive skyscraper. Mathematical discovery takes place in the dark, functioning like a narrow shaft of light that gradually traces out the structure's framework by illuminating the connections from one steel girder to the next. As we see one piece after another tied together, our confidence in the structure grows, and we develop a unified picture of mathematics.

But we could be missing a neighboring structure, no matter how close, if it has no direct links with the one we're working on. In our logical pursuit, passing from one connected girder to another, we would never come across the alternate paths that closely parallel our own. There's really no reason to suppose that two different formulations of mathematics, intertwined but independent, couldn't develop. That would make extraterrestrial communication very tricky.

Mathematician David Ruelle suggests that in a few decades, we may get a chance to see what at least one type of nonhuman mathematics would look like. "I am not predicting the imminent arrival of little green men from outer space, but simply the invasion of mathematics by computers," he says "My guess is that, within 50 or 100 years (or it might be 150), computers will successfully compete with the human brain in doing mathematics, and that their mathematical style will be rather different from ours. Fairly long computational verifications (numerical or combinatorial) will not bother them at all, and this should lead not just to different sorts of proofs, but more importantly to different sorts of theorems being proved."

Figures of Beauty

In the words of mathematician and philosopher Bertrand Russell, "Mathematics, rightly viewed, possesses not only truth, but supreme beauty—a beauty cold and austere, like that of sculpture, without appeal to any part of our weaker nature, without the trappings of paintings or music, yet sublimely pure, and capable of stern perfection such as only the greatest art can show."

Beauty is a culturally acquired taste. To appreciate mathematical beauty requires long education and training. Although beauty is basically a subjective perception, mathematicians agree to a surprising degree on what is beautiful and what is not. Thus, mathematicians strive not just to construct irrefutable proofs but also to present their ideas and results in a clear and compelling fashion, dictated more by a sense of aesthetics than by the needs of logic. They are concerned not merely with finding and proving theorems but also with arranging and assembling the theorems into an elegant, coherent structure.

Considerations of beauty also figure in the choice of problems to pursue and methods used to prove conjectures. Mathematician Morris Kline has said, "Much research for new proofs of theorems already correctly established is undertaken simply because the existing proofs have no aesthetic appeal."

Beauty and power in mathematics are strongly correlated. Mathematicians find concepts and structures beautiful if those notions enable them to derive new results, see new patterns, understand new phenomena, and establish new links between diverse parts of mathematics. Henri Poincaré noted: "It is true aesthetic feeling which all mathematicians recognize . . . The useful combinations are precisely the most beautiful."

The lure of beauty is so strong that when faced with a messy truth and sublime beauty, mathematicians tend to choose beauty. Mathematician Hermann Weyl once said: "My work has always tried to unite the true with the beautiful, but when I choose one over the other, I usually choose the beautiful." Mathematician G. H. Hardy put it another way: "Beauty is the first test: There is no permanent place in the world for ugly mathematics."

This sense of mathematical beauty is foreign to most outsiders. Without a highly educated eye, it's hard to look at a proof or an equation and remark on its beauty. But nonmathematicians can appreciate some types of mathematical beauty, especially those found in geometric structures and patterns, which appeal to the visual sense.

Unfortunately, during most of the twentieth century, mathematicians frowned upon the use of diagrams in their expositions and arguments. Even with problems so unavoidably visual as tiling a plane, proofs of solutions have more often than not consisted of rows of symbols rather than diagrams. Each row follows from the previous row in accordance with the laws of mathematical logic. There's been little room in mathematical discourse for diagrams or arguments that appeal to common sense or intuition.

Now, especially with the advent of computer graphics, professional mathematicians are rediscovering the advantages of pictorial reasoning and argument, even in problems having no obvious connection with geometry. Thus, computer graphics has added a new element to the beauty of mathematics. Over the last few years, mathematicians have begun to explore and enjoy patterns in their equations and other mathematical formulations, made visible by computer graphics. Using computer-based techniques, they have discovered graceful geometric forms reminiscent of soap-film surfaces; studied the bizarre, chaotic results of iterating simple equations (see Color Plates 15 and 16); visualized higher dimensions; and penetrated the infinitely detailed world of fractals.

Computer graphics also allows nonmathematicians to experience some of the pleasure that mathematicians take in their work. While mathematicians use such images to inspire and further their research, nonmathematicians are able to appreciate some of the mathematical qualities portrayed in the pictures. In fact, anyone with a little imagination, some skill in writing computer programs, and access to a computer can now generate breathtaking images of mathematical objects (see Figure 8.7).

To appreciate more of the austere beauty of mathematics requires much effort and dedication. In 1930, German mathema-

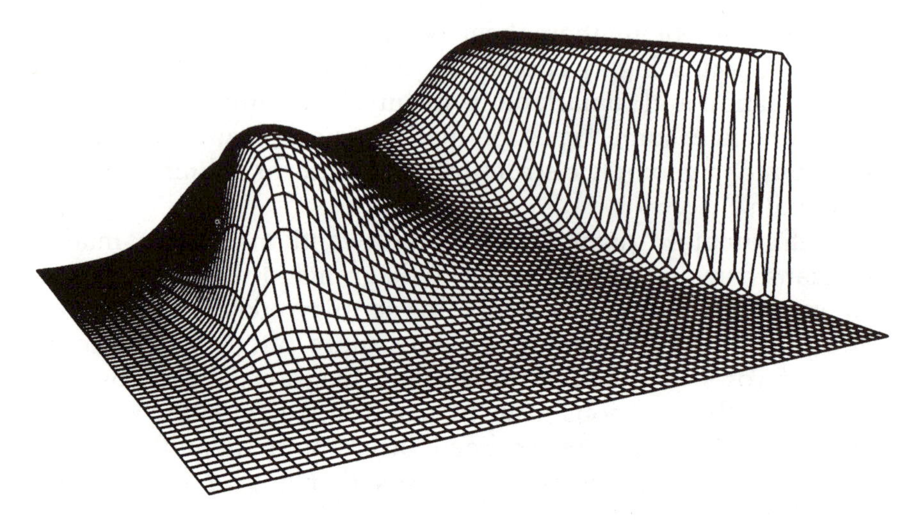

Figure 8.6 This computer-generated image illustrates an equation describing a surface with one local maximum, or peak. Changing the scale and viewpoint reveals that this function, $3xe^y - x^3 - e^{3y}$, has a number of interesting properties.

tician Wolfgang Krull reflected on the isolation mathematicians sometimes feel:

> The more we ourselves are enraptured by the beauties of mathematics, the more we regret that we can bring so few people to share our pleasure. But at least those of us in the school of abstract mathematics have one consolation: as we make our presentations clearer and more transparent, they automatically become easier to understand. Bear in mind that four hundred years ago, arithmetic was a difficult art. So great an educator as Melanchthon [a sixteenth-century scholar who reformed German education] did not trust the average student to penetrate the secrets of fractions. Yet now every child in elementary school must master them. Perhaps eventually the beauties of higher mathematics . . . will be accessible to every educated person.

Figure 8.7 Mathematicians have developed a variety of techniques for solving equations of the form $f(z) = 0$. The choice of method influences how long it takes to arrive at a solution and, if the equation has several solutions, which one turns up first. This figure gives an indication of how well Halley's method works when applied to the equation $(z^2 - 1) = 7$, which has seven solutions. The black regions encompass starting points that rapidly converge to any one of the equation's solutions. Such plots show when the method can be relied upon and where it behaves strangely.

In their search for patterns and logical connections, mathematicians face a vast, mysterious ocean of possibilities. Over the centuries, they have discovered an extensive archipelago of truth and beauty. Much of that accumulated knowledge is passed on to succeeding generations. Even more wonders await future explorers of deep, mathematical waters.

Further Reading

1 *Beginnings*

"An Ancient Token System: The Precursor to Numerals and Writing," Denise Schmandt-Besserat in *Archaeology*, Vol. 39, November–December 1986, pages 32–39.

"Tokens: Facts and Interpretation," Denise Schmandt-Besserat in *Visible Language*, Vol. 20, No. 3 (Summer 1986), pages 250–273.

"Oneness, Twoness, Threeness," Denise Schmandt-Besserat in *The Sciences*, Vol. 27 (1987), No. 4, pages 44–48.

A History of Mathematics. Carl B. Boyer. Princeton University Press, 1985.

"The Science of Patterns," Lynn Arthur Steen in *Science*, Vol. 240, 29 April 1988, pages 611–616.

Mathematics Today. Lynn Arthur Steen, editor. Springer-Verlag, 1978.

Mathematical Sciences: Some Research Trends. Board on Mathematical Sciences, National Research Council. National Academy Press, 1988.

Number: The Language of Science (Fourth Edition). Tobias Dantzig. The Free Press, 1954.

What is Mathematics? Richard Courant and Herbert Robbins. Oxford University Press, 1941.

Mathematics and the Imagination. Edward Kasner and James Newman. Microsoft Press, 1989.

Mathematics: The New Golden Age. Keith Devlin. Penguin Books, 1988.

"Constructing Crystalline Minimal Surfaces," Jean E. Taylor in *Seminar on Minimal Submanifolds*. Princeton University Press, 1983.

"Computer Images in Five Dimensions," William J. Cromie in *Mosaic*, Vol. 19, No. 2 (Summer 1988), pages 16–31.

Mathematics and Optimal Form. Stefan Hildebrandt and Anthony Tromba. Scientific American Books, 1985.

"Computer Graphics Tools for the Study of Minimal Surfaces," Michael J. Callahan, David Hoffman, and James T. Hoffman in *Communications of the ACM*, Vol. 31, June 1988, pages 648–661.

"Mathematicians at the Receiving End," T. A. Heppenheimer in *Mosaic*, Vol. 16, No. 4 (Winter 1986), pages 37–47.

"The Mathematics of Three-Dimensional Manifolds," William P. Thurston and Jeffrey R. Weeks in *Scientific American*, Vol. 251, July 1984, pages 108–120.

"Three-Dimensional Manifolds, Kleinian Groups and Hyperbolic Geometry," William P. Thurston in *Bulletin of the American Mathematical Society*, Vol. 6, No. 3 (May 1982), pages 357–381.

"The Unreasonable Effectiveness of Mathematics in the Natural Sciences," Eugene P. Wigner in *Communications on Pure and Applied Mathematics*, Vol. 13 (1960), pages 1–14.

"Physics and Mathematics at the Frontier," David J. Gross in *Proceedings of the National Academy of Sciences (USA)*, Vol. 85, November 1988, pages 8371–8375.

2 *New Twists*

Mathematische Modelle/Mathematical Models. Gerd Fischer, editor. Vieweg, 1986.

A Topological Picturebook. George K. Francis. Springer-Verlag, 1987.

"The Classification of Surfaces," Peter Andrews in *The American Mathematical Monthly*, Vol. 95, November 1988, pages 861–868.

Surface Topology. P. A. Firby and C. F. Gardiner. Ellis Horwood, 1982.

The Mathematical Description of Space and Form. E. A. Lord and C. B. Wilson. Ellis Horwood, 1986.

"Using Supercomputers to Visualize Higher Dimensions: An Artist's Contribution to Scientific Visualization," Donna J. Cox in *Leonardo,* Vol. 21 (1988), No. 3, pages 233–242.

"Turning a Surface Inside Out," Anthony Phillips in *Scientific American,* Vol. 214, May 1966, pages 112–120.

"Arnold Shapiro's Eversion of the Sphere," George K. Francis and Bernard Morin in *The Mathematical Intelligencer,* Vol. 2 (1979), No. 4, pages 200–203.

"The New Polynomial Invariants of Knots and Links," W. B. R. Lickorish and K. C. Millett in *Mathematics Magazine,* Vol. 61, February 1988, pages 3–23.

"New Invariants in the Theory of Knots," Louis H. Kauffman in *The American Mathematical Monthly,* Vol. 95, No. 3 (March 1988), pages 195–242.

"Topologists Startled by New Results," Gina Kolata in *Science,* Vol. 217, 30 July 1982, pagaes 432–433.

"The Mathematics of Manifolds," T. A. Heppenheimer in *Mosaic,* Vol. 19, No. 2 (Summer 1988), pages 32–43.

"Instantons and the Topology of 4-Manifolds," Ronald J. Stern in *The Mathematical Intelligencer,* Vol. 5 (1983), No. 3, pages 39–44.

"Mysteries of four dimensions," John D. S. Jones in *Nature,* Vol. 332, 7 April 1988, pages 488–489.

"Mathematics of a Fake World," Andrew Watson in *New Scientist,* Vol. 118, 2 June 1988, pages 41–45.

3 *Fitting Arrangements*

Tilings and Patterns. Branko Grünbaum and G. C. Shephard. W. H. Freeman, 1987.

"Tiling with Convex Polygons" in *Time Travel and Other Mathematical Bewilderments.* Martin Gardner. W. H. Freeman, 1988.

"Picture Puzzling," Ivan Rival in *The Sciences*, Vol. 27, January-February 1987, pages 40–46.

"In Praise of Amateurs," Doris Schattschneider in *The Mathematical Gardner*. David A. Klarner, editor. Weber and Schmidt, 1981.

For All Practical Purposes: Introduction to Contemporary Mathematics. Project director, Solomon Garfunkel; coordinating editor, Lynn A. Steen. W. H. Freeman, 1988.

Penrose Tiles to Trapdoor Ciphers. Martin Gardner. W. H. Freeman, 1989.

"Opening the Door to Forbidden Symmetries," Mort La Brecque in *Mosaic*, Vol. 18, No. 4 (Winter 1987/1988), pages 2–23.

"Quasicrystals," David R. Nelson in *Scientific American*, Vol. 255, August 1986, pages 42–51.

"Growing Perfect Quasicrystals," George Y. Onoda, Paul J. Steinhardt, David P. DiVincenzo, and Joshua E. S. Socolar in *Physical Review Letters*, Vol. 60, 20 June 1988, pages 2653–2656.

"Theory of Matching Rules for the 3-Dimensional Penrose Tilings," A. Katz in *Communications in Mathematical Physics*, Vol. 118, August 1988, Pages 263–288.

"The Packing of Spheres," N. J. A. Sloane in *Scientific American*, Vol. 250, January 1984, pages 116–125.

"Sphere Packing," François Sigrist in *The Mathematical Intelligencer*, Vol. 5 (1983), No. 3, pages 34–38.

Sphere Packings, Lattices and Groups. J. H. Conway and N. J. A. Sloane. Springer-Verlag, 1988.

The Architecture Machine: Toward a More Human Environment. Nicholas Negroponte. MIT Press, 1970.

The Geometry of Environment: An Introduction to Spatial Organization in Design. Lionel March and Philip Steadman. MIT Press, 1974.

Leon Battista Alberti. Franco Borsi. Harper & Row, 1977.

"Spatial Systems in Architecture and Design: Some History and Logic," L. March and G. Stiny in *Environment and Planning B: Planning and Design*, Vol. 12, February 1985, pages 31–53.

4 *Snowflake Curves*

The Fractal Geometry of Nature. Benoit B. Mandelbrot. W. H. Freeman, 1982.

The Science of Fractal Images. Heinz-Otto Peitgen and Dietmar Saupe, editors. Springer-Verlag, 1988.

Fractals Everywhere. Michael Barnsley. Academic Press, 1988.

Fractals. Jens Feder. Plenum Press, 1988.

"Real Time Design and Animation of Fractal Plants and Trees," Peter E. Oppenheimer in *Proceedings of Siggraph '86,* Vol. 20 (1986), No. 4, pages 55–64.

"Images Generated by Orbits of 2-D Markov Chains," Marc A. Berger in *Chance,* Vol. 2 (1989), No. 2, pages 18–28.

"Fractals: Not Just Another Pretty Picture," Glenn Zorpette in *IEEE Spectrum,* Vol. 25, October 1988, pages 29–31.

Exploring the Geometry of Nature: Computer Modeling of Chaos, Fractals, Cellular Automata and Neural Networks. Edward Rietman. Windcrest Books, 1989.

"The Growth of Snow Crystals," B. J. Mason in *Scientific American,* Vol. 204, January 1961, pages 120–131.

"Snow Crystals," Charles and Nancy Knight in *Scientific American,* Vol. 228, January 1973, pages 100–107.

"Fractal Concepts in Physics," Fereydoon Family in *Naval Research Reviews,* Vol. 40, Four/1988-One/1989, pages 2–13.

"Dendrites, Viscous Fingers, and the Theory of Pattern Formation," J. S. Langer in *Science,* Vol. 243, 3 March 1989, pages 1150–1156.

"Tip Splitting Without Interfacial Tension and Dendritic Growth Patterns Arising from Molecular Anisotropy," Johann Nittmann and H. Eugene Stanley in *Nature,* Vol. 321, 12 June 1986, pages 663–668.

"Role of Fluctuations in Fluid Mechanics and Dendritic Solidification," H. Eugene Stanley in *Philosophical Magazine B,* Vol. 56, No. 2, 1987, pages 665–686.

"Non-deterministic Approach to Anisotropic Growth Patterns with Continuously Tunable Morphology: The Fractal Properties of Real Snowflakes," Johann Nittmann and H. Eugene Stanley in *Journal of Physics A*, Vol. 20 (1987), pages L1185–L1191.

"Deterministic Growth Model of Pattern Formation in Dendritic Solidification," Fereydoon Family, Daniel E. Platt, and Tamás Vicsek in *Journal of Physics A*, Vol. 20 (1987), pages L1177–L1183.

"On the Williams-Watts Function of Dielectric Relaxation," Michael F. Shlesinger and Elliott W. Montroll in *Proceedings of the National Academy of Science (USA)*, Vol. 81, February 1984, pages 1280–1283.

"Generalized Vogel Law for Glass-Forming Liquids," John T. Bendler and Michael F. Shlesinger in *Journal of Statistical Physics*, Vol. 53 (1988), Nos. 1 and 2, pages 531–541.

"Self-Organized Criticality," Per Bak, Chao Tang, and Kurt Wiesenfeld in *Physical Review A*, Vol. 38 (1988), pages 364–374.

"Self-Organized Criticality: An Explanation of 1/f Noise," Per Bak, Chao Tang, and Kurt Wiesenfeld in *Physical Review Letters*, Vol. 59, 27 July 1987, pages 381–384.

"Properties of Earthquakes Generated by Fault Dynamics," J. M. Carlson and J. S. Langer in *Physical Review Letters*, Vol. 62, 29 May 1989, pages 2632–2635.

"Relaxation at the Angle of Repose," H. M. Jaeger, Chu-heng Liu, and Sidney R. Nagel in *Physical Review Letters*, Vol. 62, 2 January 1989, pages 40–43.

"White and Brown Music, Fractal Curves and One-over-f Fluctuations," Martin Gardner in *Scientific American*, Vol. 238, April 1978, pages 16–32.

5 *Number Play*

"The Strong Law of Small Numbers," Richard K. Guy in *The American Mathematical Monthly*, Vol. 95, October 1988, pages 697–712.

"Alphamagic Squares," Lee C. F. Sallows in *Abacus*, Vol. 4, No. 1 (Fall 1986), pages 28–45.

"Alphamagic Squares, Part II," Lee C. F. Sallows in *Abacus*, Vol. 4, No. 2 (Winter 1987), pages 20–29, 43.

"Fourteen Proofs of a Result About Tiling a Rectangle," Stan Wagon in *The American Mathematical Monthly*, Vol. 94, August–September 1987, pages 601–617.

"Polyominoes," in *Mathematical Puzzles & Diversions*. Martin Gardner. Simon and Schuster, 1959.

"Tiling with Polyominoes," Solomon W. Golomb in *Journal of Combinatorial Theory*, Vol. 1, September 1966, pages 280–296.

"Tiling with Sets of Polyominoes," Solomon W. Golomb in *Journal of Combinatorial Theory*, Vol. 9, July 1970, pages 60–71.

"Polyominoes Which Tile Rectangles," Solomon W. Golomb in *Journal of Combinatorial Theory A*, Vol. 51, May 1989, pages 117–124.

"The Y-Hexomino Has Order 92," Karl A. Dahlke in *Journal of Combinatorial Theory A*, Vol. 51, May 1989, pages 125–126.

"A Heptomino of Order 76," Karl A. Dahlke in *Journal of Combinatorial Theory A*, Vol. 51, May 1989, pages 127–128.

"Are the Twin Circle of Archimedes Really Twins?" Leon Bankoff in *Mathematics Magazine*, Vol. 47, September-October 1974, pages 214–218.

"Ramanujan—100 Years Old (Fashioned) or 100 Years New (Fangled)?" Bruce C. Berndt in *The Mathematical Intelligencer*, Vol. 10 (1988), No. 3, pages 24–29.

"A Pilgrimage," Bruce C. Berndt in *The Mathematical Intelligencer*, Vol. 8 (1986), No. 1, pages 25–30.

Ramanujan Revisited. George E. Andrews, Richard A. Askey, Bruce C. Berndt, K. G. Ramanathan, and Robert A. Rankin, editors. Academic Press, 1988.

"The Formula Man," Ian Stewart in *New Scientist*, Vol. 116, 17 December 1987, pages 24–28.

"Questions and Conjectures in Partition Theory," George E. Andrews in *The American Mathematical Monthly*, Vol. 93, November 1986, pages 708–711.

"Ramanujan and Pi," Jonathan M. Borwein and Peter B. Borwein in *Scientific American*, Vol. 258, February 1988, pages 112–117.

"Is π Normal?" Stan Wagon in *The Mathematical Intelligencer*, Vol. 7 (1985), No. 3, pages 65–67.

"The Computation of π to 29,360,000 Decimal Digits Using Borweins' Quartically Convergent Algorithm," David H. Bailey in *Mathematics of Computation*, Vol. 50, January 1988, pages 283–296

"Pi: Difficult or Easy?" G. A. Edgar in *Mathematical Magazine*, Vol. 60, June 1987, pages 141–150.

"The Ubiquitous π," Dario Castellanos in *Mathematics Magazine*, Vol. 61, April 1988, pages 67–98.

"Ramanujan, Modular Equations, and Approximations to Pi or How to Compute One Billion Digits of Pi," J. M. Borwein, P. B. Borwein, and D. H. Bailey in *The American Mathematical Monthly*, Vol. 96, March 1989, pages 201–219.

Number Theory in Science and Communication. M. R. Schroeder. Springer-Verlag, 1984.

"Number Theory and the Real World," Manfred R. Schroeder in *The Mathematical Intelligencer*, Vol. 7 (1985), No. 4, pages 18–26.

"Toward Better Acoustics for Concert Halls," Manfred R. Schroeder in *Physics Today*, Vol. 33, October 1980, pages 24–30.

"The Reflection Phase Grating Acoustical Diffusor: Diffuse It Or Lose It," Peter D'Antonio in *db*, Vol. 19, September–October 1985, pages 46–49.

6 *Hard Times*

Labyrinths of Reason. William Poundstone. Anchor Press, 1988.

"Infeasible Computation: NP-Complete Problems," Edmund A. Lamagna in *Abacus*, Vol. 4, No. 3 (Spring 1987), pages 18–33.

The Turning Omnibus: 61 Excursions in Computer Science. A. K. Dewdney. Computer Science Press, 1989.

Algorithmics: The Spirit of Computing. David Harel. Addison-Wesley, 1987.

"Aspects of the Traveling Salesman Problem," M. Held, A. J. Hoffman, E. L. Johnson, and P. Wolfe in *IBM Journal of Research and Development*, Vol. 28, July 1984, pages 476–486.

"Intrinsically Difficult Problems," Larry J. Stockmeyer and Ashok K. Chandra in *Scientific American*, Vol. 240, May 1979, pages 140–159.

"The Mathematician as an Explorer," Sherman K. Stein in *Scientific American*, Vol. 204, May 1961, pages 148–158.

"Must 'Hard Problems' Be Hard?" Gina Kolata in *Science*, Vol. 228, 26 April 1985, pages 479–481.

"The Shortest-Network Problem," Marshall W. Bern and Ronald L. Graham in *Scientific American*, Vol. 260, January 1989, pages 84–89.

"Combinatorics, Complexity, and Randomness," Richard M. Karp in *Communications of the ACM*, Vol. 29, February 1986, pages 98–111.

"An Analogue Approach to the Travelling Salesman Problem Using an Elastic Net Method," Richard Durbin and David Willshaw in *Nature*, Vol. 326, 16 April 1987, pages 689–691.

"Optimization by Simulated Annealing," S. Kirkpatrick, C. D. Gelatt, Jr., and M. P. Vecchi in *Science*, Vol. 220, 13 May 1983, pages 671–680.

"Image Processing by Simulated Annealing," P. Carnevali, L. Coletti, and S. Patarnello in *IBM Journal of Research and Development*, Vol. 29, November 1985, pages 569–579.

"New Optimization Methods from Physics and Biology," David G. Bounds in *Nature*, Vol. 329, 17 September 1987, pages 215–219.

"Genetic Algorithms: What Computers Can Learn from Darwin," Charles T. Walbridge in *Technology Review*, Vol. 92, January 1989, pages 47–53.

Genetic Algorithms in Search, Optimization, and Machine Learning. David E. Goldberg. Addison-Wesley, 1989.

"The Several Sorts of Sorts," Hugh Kenner in *Discover*, Vol. 7, May 1986, pages 78–83.

"Algorithm Design," Robert E. Tarjan in *Communications of the ACM*, Vol. 30, March 1987, pages 204–212.

"Chess Computers Make Their Move," Greg Wilson in *New Scientist*, Vol. 123, 5 August 1989, pages 50–53.

Computer Chess. David E. Welsh. Wm. C. Brown, 1983.

Computer Gamesmanship. David Levy. Simon & Schuster, 1983.

"Chess Master Versus Computer," David Levy in *Abacus*, Vol. 2 (1985), No. 2, pages 72–77.

"Computer Chess and the Humanization of Technology," Donald Michie in *Nature*, Vol. 299, 30 September 1982, pages 391–394.

"Machine Beats Man on Ancient Front," James Gleick in *The New York Times*, August 26, 1986, pages C1, C8.

"An Advice-Taking Chess Computer," Albert L. Zobrist and Frederic R. Carlson, Jr., in *Scientific American*, Vol. 228, June 1973, pages 92–105.

"Using Chunking to Play Chess Pawn Endgames," Murray Campbell and Hans Berliner in *Artificial Intelligence*, Vol. 23, May 1984, pages 97–120.

7 *Shadows of Chaos*

Mathematics and the Unexpected. Ivar Ekeland. The University of Chicago Press, 1988.

Chaos: The Making of a New Science. James Gleick. Viking, 1987.

"Randomness of a True Coin Toss," Vladimir Z. Vulović and Richard E. Prange in *Physical Review A*, Vol. 33, January 1986, pages 576–582.

The Eudaemonic Pie. Thomas A. Bass. Houghton Mifflin, 1985.

"Prestidigitator of Digits," Gina Kolata in *Science 85*, Vol. 6, April 1985, pages 67–72.

"Particles in Motion: The Case of the Loaded Die," Bradley T. Werner in *Engineering & Science*, Vol. 49, March 1986, pages 20–24.

"Fair Dice," Persi Diaconis and Joseph B. Keller in *The American Mathematical Monthly*, Vol. 96, April 1989, pages 337–339.

"Shuffling Cards and Stopping Times," David Aldous and Persi Diaconis in *The American Mathematical Monthly*, Vol. 93, May 1986, pages 333–348.

"What Does It Mean to Be Random?" Gina Kolata in *Science*, Vol. 231, 7 March 1986, pages 1068–1070.

"A Current View of Random Number Generators," George Marsaglia in *Computer Science and Statistics: The Interface*. L. Billard, editor. Elsevier Science Publishers, 1985.

"Random Numbers Fall Mainly in the Planes," George Marsaglia in *Proceedings of the National Academy of Sciences (USA)*, Vol. 61, September 1968, pages 25–28.

"The Orderly Pursuit of Pure Disorder," Kevin McKean in *Discover*, Vol. 8, January 1987, pages 72–81.

"A Simple Unpredictable Pseudo-Random Number Generator," L. Blum, M. Blum, and M. Shub in *SIAM Journal of Computation*, Vol. 15, May 1986, pages 364–383.

Chaotic Vibrations. Francis C. Moon. John Wiley & Sons, 1987.

"Chaotic Dynamics of a Bouncing Ball," N. B. Tufillaro and A. M. Albano in *American Journal of Physics*, Vol. 54, October 1986, pages 939–944.

"Chaos in the Swing of a Pendulum," David Tritton in *New Scientist*, Vol. 114, 24 July 1986, pages 37–40.

"Chaos, Strange Attractors, and Fractal Basin Boundaries in Nonlinear Dynamics," Celso Grebogi, Edward Ott, and James A. Yorke in *Science*, Vol. 238, 30 October 1987, pages 632–638.

"Chaos," James P. Crutchfield, J. Doyne Farmer, Norman H. Packard, and Robert S. Shaw in *Scientific American*, Vol. 255, December 1986, pages 46–57.

"Nonlinear and Chaotic String Vibrations," Nicholas B. Tufillaro in *American Journal of Physics*, Vol. 57, May 1989, pages 408–414.

"Computer-Drawn Pictures Stalk the Wild Trajectory," Barry A. Cipra in *Science*, Vol. 241, 2 September 1988, pages 1162–1163.

"Numerical Orbits of Chaotic Processes Represent True Orbits," Stephen M. Hammel, James A. Yorke, and Celso Grebogi in *Bulletin of the American Mathematical Society*, Vol. 19, October 1988, pge 465–469.

The Beauty of Fractals. H.-O. Peitgen and P. H. Richter. Springer-Verlag, 1986.

Does God Play Dice? Ian Stewart. Basil Blackwell, 1989.

"Numerical Evidence That the Motion of Pluto is Chaotic," Gerald Jay Sussman and Jack Wisdom in *Science*, Vol. 241, 22 July 1988, pages 433–437.

"Playing Dice with the Solar System," Anita M. Killian in *Sky & Telescope*, Vol. 78, August 1989, pages 136–140.

"Chaos Theory: How Big an Advance?" Robert Pool in *Science*, Vol. 245, 7 July 1989, pages 26–28.

"Ergodic Theory, Randomness,and 'Chaos'," D. S. Ornstein in *Science*, Vol. 243, 13 January 1989, pages 182–187.

"Laboratory Simulation of Jupiter's Great Red Spot," Joel Sommeria, Steven D. Meyers, and Harry L. Swinney in *Nature*, Vol. 331, 25 February 1988, pages 689–693.

"Numerical Simulation of Jupiter's Great Red Spot," Philip S. Marcus in *Nature*, Vol. 331, 25 February 1988, pages 693–696.

"How Random is a Coin Toss?" Joseph Ford in *Physics Today*, Vol. 36, April 1983, pages 40–47?

"A New Source of Complexity," Paul Davies in *New Scientist*, Vol. 120, 26 November 1988, pages 48–50.

"Randomness and Mathematical Proof," Gregory J. Chaitin in *Scientific American*, Vol. 232, May 1975, pages 47–52.

"The Ultimate in Undecidability," Ian Stewart in *Nature*, Vol. 332, 10 March 1988, pages 115–116.

"Randomness in Arithmetic," Gregory J. Chaitin in *Scientific American*, Vol. 259, July 1988, pages 80–85.

"Master of the Incomplete," Rudy Rucker in *Science 82*, Vol. 3, October 1982, pages 56–60.

"The Incompleteness of Arithmetic," Arturo Sangelli in *New Scientist*, Vol. 116, 5 November 1987, pages 42–45.

"Directions in Classical Chaos," Joseph Ford in *Directions in Chaos*. Hao Bai-Lin, editor. World Scientific Publishing, 1988.

8 Truth and Beauty

"Computer Search Solves an Old Math Problem," Barry A. Cipra in *Science*, Vol. 242, 16 December 1988, pages 1507–1508.

"The Circle Can Be Squared!" Barry A. Cipra in *Science*, Vol. 244, 5 May 1989, page 528.

"Hungarian Mathematician Squares the Circle," Barry A. Cipra in *SIAM News*, September 1989, page 19.

"The Solution of the Four-Color-Map Problem," Kenneth Appel and Wolfgang Haken in *Scientific American*, Vol. 237, October 1977, pages 108–121.

"Do Mathematicians Still Do Math?" Barry A. Cipra in *Science*, Vol. 244, 19 May 1989, pages 769–770.

"The Enormous Theorem," Daniel Gorenstein in *Scientific American*, Vol. 253, December 1985, pages 104–115.

"Demystifying the Monster," Ian Stewart in *Nature*, Vol. 319, 20 February 1986, pages 621–622.

"Ten Thousand Pages to Prove Simplicity," Mark Cartwright in *New Scientist*, Vol. 109, 30 May 1985, pages 26–30.

"Fermat's Last Theorem," Harold M. Edwards in *Scientific American*, Vol. 239, October 1978, pages 104–122.

"The First Case of Fermat's Last Theorem," D. R. Heath-Brown in *The Mathematical Intelligencer*, Vol. 7 (1985), No. 4, pages 40–47, 55.

"Progress on Fermat's Famous Math Problem," Gina Kolata in *Science*, Vol. 235, 27 March 1987, pages 1572–1573.

"The Last Theorem," Ian Stewart in *New Scientist*, Vol. 118, 21 April 1988, pages 64–65.

"Fermat's Last Theorem Remains Unproved," Barry A. Cipra in *Science*, Vol. 240, 3 June 1988, pages 1275–1276.

"Fermat Still Has Last Laugh," Paul Hoffman in *Discover*, Vol. 10, January 1989, pages 48–50.

"Is Our Mathematics Natural? The Case of Equilibrium Statistical Mechanics," David Ruelle in *Bulletin of the American Mathematical Society*, Vol. 19, July 1988, pages 259–268.

"Constructive Mathematics," Allan Calder in *Scientific American*, Vol. 241, October 1979, pages 146–171.

"The Centrality of Mathematics in the History of Western Thought," Judith V. Grabiner in *Mathematics Magazine*, Vol. 61, October 1988, pages 220–230.

"The Truth and Nothing But the Truth," David Gale in *The Mathematical Inttelligencer*, Vol. 11 (1989), No. 3, pages 62–67.

"Unsolved Problems in Arithmetic," Howard DeLong in *Scientific American*, Vol. 224, March 1971, pages 50–60.

"The Aesthetic Viewpoint in Mathematics," Wolfgang Krull (translated by Betty S. and William C. Waterhouse) in *The Mathematical Intelligencer*, Vol. 9 (1987), No. 1, pages 48–52.

"Catch of the Day: Biomorphs on Truchet Tiles, Served with Popcorn and Snails," A. K. Dewdney in *Scientific American*, Vol. 261, July 1989, pages 110–113.

"Pattern Formation and Chaos in Networks," Clifford A. Pickover in *Communications of the ACM*, Vol. 31, February 1988, pages 136–151.

"Overrelaxation and Chaos," Clifford A. Pickover in *Physics Letters A*, Vol. 130, 4 July 1988, pages 125–128.

Sources of Illustrations

Pages 4, 5
Musée du Louvre, Département des antiquités orientales, Paris, France

Page 7
Royal Ontario Museum, Godin Project

Pages 19, 20
Jean E. Taylor, Rutgers University

Pages 22 (Figure 1.9, *top*), **23** (Figure 1.10, *bottom*), **Color Plate 1**
© James T. Hoffman, GANG, UMass Amherst

Pages 22 (Figure 1.9, *bottom*), **23** (Figure 1.10, *top and middle*)
E. Thomas, J. Hoffman, D. Anderson, and C. Henkee

Page 25
David Dobkin, Princeton University

Page 27
Jeff Weeks

Pages 28, 289
Stan Wagon, *Doing Mathematics with Mathematica*, W. H. Freeman and Company

Page 30
Charles W. Misner, Kip S. Thorne, and John Archibald Wheeler, *Gravitation*, © 1973 by W. H. Freeman and Company

Page 34
Gerd Fischer (editor), *Mathematische Modelle/Mathematical Models* (Vol. 1–2), Friedr. Vieweg & Sohn, 1986

Pages 38, 63, 81, 201
Lynn Arthur Steen (editor), *Mathematics Today*, Springer-Verlag, © 1978 by the Conference Board of the Mathematical Sciences

Pages 39, 45, Color Plates 2, 3
Donna Cox, George Francis, and Ray Idaszak, NCSA

Page 41
George K. Francis, *A Topological Picturebook*, Springer-Verlag, 1987

Pages 42, 164, 165
Stan Wagon

Pages 47, 49, 53
Irving Geis and Alan Beechel, *Scientific American*, May 1966, pages 113, 116

Page 50
Bernard Morin, Université Louis Pasteur, Strasbourg, France

Page 69
National Optical Astronomy Observatories

Pages 72, 74, 78, 85 (Figure 3.11), **87** (Figure 3.13, *right*)
Branko Grünbaum and G. C. Shephard, *Tilings and Patterns*, © 1987 by W. H. Freeman and Company

Page 75
James Egleson, *Scientific American*, April 1961, p. 170

Pages 83, 84, 85 (Figure 3.12)
Martin Gardner, *Time Travel and Other Mathematical Bewilderments*, © 1988 by W. H. Freeman and Company

Page 87 (Figure 3.13, *left*)
A. I. Goldman and P. W. Stephens

Pages 89, 90, 93, 94, Color Plate 8
David DiVincenzo, Steven Langer, and Kevin Ingersent

Page 98
N. J. A. Sloane, Bell Laboratories, and Gabor Kiss, *Scientific American*, January 1984, page 121

Pages 100, 101
Gabor Kiss, *Scientific American*, January 1984, pages 118, 120

Page 105
Illustration from *Leon Battista Alberti* by Franco Borsi. English translation copyright © 1975 by Electra Editrice. Reprinted by permission of Harper & Row, Publishers, Inc., and by courtesy of Elemond, Milano.

Page 106
George Stiny and Lionel March

Page 108
All drawings of work by Frank Lloyd Wright © 1990 FLWright Fdn

Pages 113, 120, 121
Benoit B. Mandelbrot, *The Fractal Geometry of Nature*, W. H. Freeman and Company, © 1982 by Benoit B. Mandelbrot

Pages 116, 122, 123, 124
Philip T. Hodge

Page 117
Martin Gardner, *Penrose Tiles to Trapdoor Ciphers*, © 1989 by W. H. Freeman and Company

Page 118
Alex Pentland, MIT Media Lab

Page 126
W. A. Bentley and W. J. Humphries, *Snow Crystals*, Dover Publications, 1962.

Pages 129, 132 (Figure 4.13, *bottom*), **Color Plate 10**
Fereydoon Family, Emory University

Page 130
Jim Egleson, *Scientific American*, January 1973, page 103

Page 132 (Figure 4.13, *top*)
Fereydoon Family, Daniel Platt, and Tamás Vicsek, *Journal of Physics A20*, L1177 (1987)

Page 138
John Bendler, GE Research & Development Center

Pages 141, 142, 143
P. Bak, C. Tang, K. Wiesenfeld, *Phys. Rev. A 38*, 364 (1989)

Page 145
Richard K. Voss, *Scientific American*, April 1978, page 21

Page 147
P. Bak, Kan Chen, C. Tang

Page 149
P. Bak, Kan Chen, M. Creutz

Page 163
Lee C. F. Sallows

Page 168
Solomon W. Golomb

Pages 172, 174
John Moss, courtesy of the Royal Society of London, *Scientific American*, February 1988, pages 112A, 117

Page 181
Michael Goodman, *Scientific American*, February 1988, page 117

Page 187
Tele-Image, Dallas; The Joiner-Rose Group, Dallas, (acoustical consultant); RPG Diffusor Systems, Inc.

Pages 189, 191
M. R. Schroeder, *Number Theory in Science and Communication*, Springer-Verlag, 1986

Page 199
Patricia J. Wynne, *Scientific American*, May 1979, page 140

Page 206
Copyright © 1986, Association for Computing Machinery, Inc. Reprinted by permission.

Page 207
Reprinted by permission from *Nature*, volume 326, pages 689–691. Copyright © 1987 Macmillan Magazines Ltd.

Page 210
Copyright © 1986 by International Business Machines Corporation; reprinted with permission.

Page 212
David E. Goldberg, *Genetic Algorithms,* © 1989 by Addison-Wesley Publishing Company, Inc., Reading, Mass. Page 12. Reprinted with permission of the publisher.

Pages 219, 221
Robert E. Tarjan

Page 231
Hans Berliner and Murray Campbell

Page 239
By permission of V. Ž. Vulović and R. E. Prange

Page 240
Illustration prepared by Bradley T. Werner

Page 246
George Marsaglia, Florida State University

Page 250
P. J. Holmes, "The Dynamics of Repeated Impacts with a Sinusoidally Vibrating Table," *Journal of Sound and Vibration 84,* 173–189 (1982).

Pages 252, 256, Color Plates 13, 14
Celso Grebogi, Edward Ott, Frank Varosi, and James A. Yorke, University of Maryland

Page 260
Reprinted with permission from *The Beauty of Fractals* by Heinz-Otto Peitgen and Peter H. Richter, Springer-Verlag, 1986

Page 262
J. Sommeria, S. D. Meyers, and H. L. Swinney, University of Texas at Austin

Pages 265, 266
Jerome Kuhl, *Scientific American*, May 1975, pages 48, 49

Page 277
Lorelle M. Raboni, *Scientific American*, October 1977, page 111

Page 279
Gabor Kiss, *Scientific American*, December 1985, page 112

Page 281
New York Public Library, *Scientific American*, October 1978, page 104

Page 290, Color Plate 15
Clifford A. Pickover, *Computers, Pattern, Chaos, and Beauty*, St. Martin's Press, 1990

Color Plate 4
Richard Denner

Color Plate 5
J. Hughes, Brown University; J. Vroom, D. Kamins, Stardent Computer

Color Plate 6
Thomas F. Banchoff, Brown University

Color Plate 7
Marjorie Rice

Color Plate 9
Michael Barnsley, Laurie Reuter, Arnaud Jacquin, and François Malassnet in Michael Barnsley, *Fractals Everywhere*, Academic Press, Inc., 1988.

Color Plate 11
P. Bak, Kan Chen, C. Tang, M. Cruetz

Color Plate 12
D. and G. Chudnovsky

Color Plate 16
Alan Norton, IBM Research, © IBM

Index